# INDOOR KITCHEN GARDENING

## for Beginners

**Quarto.com**

© 2025 Quarto Publishing Group USA Inc.
Text © 2014 Quarto Publishing Group USA Inc.

First Published in 2025 by New Shoe Press, an imprint of The Quarto Group,
100 Cummings Center, Suite 265-D, Beverly, MA 01915, USA.
T (978) 282-9590 F (978) 283-2742

**Essential, In-Demand Topics, Four-Color Design, Affordable Price**
New Shoe Press publishes affordable, beautifully designed books covering evergreen, in-demand subjects. With a goal to inform and inspire readers' everyday hobbies, from cooking and gardening to wellness and health to art and crafts, New Shoe titles offer the ultimate library of purposeful, how-to guidance aimed at meeting the unique needs of each reader. Reimagined and redesigned from Quarto's best-selling backlist, New Shoe books provide practical knowledge and opportunities for all DIY enthusiasts to enrich and enjoy their lives.

Visit Quarto.com/New-Shoe-Press for a complete listing of the New Shoe Press books.

New Shoe Press titles are also available at discount for retail, wholesale, promotional, and bulk purchase. For details, contact the Special Sales Manager by email at specialsales@quarto.com or by mail at The Quarto Group, Attn: Special Sales Manager, 100 Cummings Center, Suite 265-D, Beverly, MA 01915, USA.

10 9 8 7 6 5 4 3 2 1

ISBN: 978-0-7603-9788-6
eISBN: 978-0-7603-9789-3

The content in this book was previously published in *Indoor Kitchen Gardening* (Cool Springs Press 2014) by Elizabeth Millard.

Library of Congress Cataloging-in-Publication Data available

Photography: Crystal Liepa

Printed in China

# INDOOR KITCHEN GARDENING
## for Beginners

**Turn Your Home into a Year-Round Vegetable Garden**

ELIZABETH MILLARD

NEW SHOE PRESS

# Contents

# Introduction

*Growing up in Minnesota, my schools always seemed located across from cornfields and farm stands, but wearily gazing outside during math class was about as close as I got to agriculture. Although my great-grandparents and grandparents were farmers, I grew up in the suburbs, a land of uniform lawns and frozen vegetables, and although I deeply appreciated lazing around in trees and watching bees in the neighbor's garden, I never imagined I'd be spending any time digging, weeding, or talking about compost. The concept of growing food was about as foreign to me as algebra (which I also believed I'd never use).*

After a few decades in the business world, that sense of disconnection to my food remained, although I'd expanded into cooking more meals and using more than one spice at a time. It wasn't until I was in my early 40s, though, that I actually grew anything more than an appetite.

Before we first managed to get into our farmland, though, we began by growing a wealth of crops inside. Winter in Minnesota is notorious for wearing optimists down to a brittle nub, but the more experimentation we did with microgreens, pea shoots, radishes, and other tasty vegetables, the more we felt like we were extending summer into our house. As the snow buried the cars outside, we harvested wave after wave of tiny, delicious greens that kept us busy until we could find some fields to till—or at least break into the raised beds in our backyard garden in Minneapolis.

It doesn't matter if you're crammed into an urban apartment with one fern balanced on the fire escape or pondering how to use a lovely greenhouse space in your new farmhouse, anyone can use these simple tricks and techniques to develop garden abundance. Let the adventure begin.

*Elizabeth (Bossy E) and Karla (Bossy K) in a rare moment of relaxation at their Community Supported Agriculture (CSA) farm, Bossy Acres, in Minnesota.*

# PART ONE

# Growing Edibles Indoors

Living in the Midwest, I've often dreamed of growing tropical fruits in my dining room, imagining a stretch of mangoes and papayas juxtaposed against the snow-sliding-sideways view of a February afternoon. The kitchen would become a tangle of vines, lush with colorful blooms and quirky vegetables, and I'd be able to pick my breakfast while the morning coffee brews.

Theoretically, with the right conditions, I should be able to achieve at least a fraction of that daydream. For example, it's likely that I could grow a dwarf Calamondin orange tree, which is reputed to be hearty to 20° Fahrenheit, or opt for an avocado tree sprouted from a pit, waiting the four to six years it takes for the new plant to bear its own fruit.

I did attempt to achieve the somewhat impossible once, trying to grow my own loofah plant, even though I knew Minnesota is far out of the loofah's zone. I managed to get it about 5 feet tall, with luscious, broad leaves and plenty of potential, but it never fruited, only kept spinning its tendrils around curtain rods and houseplants. When the project resulted in more pruning than loofah harvesting, I came to an important realization: indoor growing is a pursuit that can be zesty and ambitious, but when it begins to feel like an overwhelming chore, it might be time to scale back. Most of all, I determined that indoor growing is best when it starts with a plan.

# Thinking Ahead

It's ridiculously easy to become overcommitted and enthusiastic, especially when perusing seed catalogs. Some of the dreariest parts of gardening—weeds, rabbits, squirrels, more weeds, birds, and did I mention weeds?—are eliminated with a kitchen counter brimming with herbs, micro-greens, lettuces, and edible flowers. So, some people tend to jump in and place a seed order that wouldn't be out of place for a five-acre hobby farm.

Before hitting "send" on that order, though, take a moment to think about what you really want to grow, and what it will add to your current growing mix (if you have one going).

Creating a plan might seem like it would take the fun out of the adventure of indoor growing, but I've found that the opposite is true. By understanding why I'm planting a specific "crop" and how I'm going to use that vegetable or herb in the future, I've been able to stay on top of my projects and very little goes to waste. I may not have tropical fruits crawling toward the ceiling, but I don't have guilt pangs from overgrown plants that have to be carted grudgingly out to the compost pile, either.

Similarly, it helps to have a strong sense of timing. Understanding when certain plants will mature and planning accordingly can be helpful for staying on top of multiple growing projects without feeling like you're now a greenhouse manager.

*Turning your home into an orangery with fresh oranges and lemons growing in abundance is a lovely dream, but you'll have better luck if you start with a more practical plan and build on your successes.*

Once you have a plan in place—even if it's a rough idea of what you want to grow—then it's easier to take a look at other factors like space, lighting, containers, etc., with a view toward creating the best conditions for your indoor growing adventures.

## WHAT'S YOUR PLAN?

For those just starting on the indoor growing path, it's much easier to choose projects/plants based on a few basic questions:

*What do you hope to gain?* If you want a crop of year-round herbs that keep your meals spicy and seasoned, think about which herbs you might use the most, and focus on those. If you're looking to boost the nutritional profile of your dishes and add flavor, consider microgreens, pea shoots, or other nutritionally dense plants that take up minimum room and deliver abundant health benefits.

*How ambitious do you want to be?* Certain plants like tomatoes, mushrooms, and potatoes aren't always easy to grow inside, but it can be done. As long as you're taking on projects like that with a sense of play and excitement, then full speed ahead, fellow grower. But if you've never grown so much as a cactus inside and suddenly want to make the leap to indoor tomatoes, you might want to add a few more baby steps into your plan. For instance, start with a reasonably sized container of herbs and once you've mastered the art of watering, pest control, and succession planting, move on to swooning over those heirloom tomato seed descriptions.

*What's your vacation schedule like?* A "quick trip" or holiday plans have derailed many of my indoor growing projects in the past. As much as I appreciate the kindness of friends who offer to water and prune, I've found it's better for me if I add vacation time into my growing plan and take a break during those times. This seems to be especially prudent when I'm growing a wide array of vegetables. No housesitter wants an elaborate 30-item list of instructions about how to deal with baby carrots, lettuces, microgreens, and mushroom bags while I'm ordering another beer from the cabana server.

*Are you looking for indoor-only growing, or transfers between the kitchen and outdoor garden?* Many gardeners extend their growing seasons by bringing some plants inside when the weather begins to cool. Herbs, in particular, are a favorite for this transfer since many can thrive fine indoors over the winter, and then go back out in the garden in the spring. If this is your goal, then that's great, but you'll need to tweak your planting mix accordingly. There are certain plants—like cilantro, for example—that simply don't do very well in making the transition. So, when choosing what to grow in your indoor space, do some research on what does best going from outdoors to indoors.

# Find Your Space

Although this book is called *Indoor Kitchen Gardening for Beginners*, there are many instances where a kitchen isn't the ideal spot in the house for vegetable or herb growth. Also, the kitchen might be perfect during a certain time of year, especially during cooler months when plants can use the ambient heat of that room, but less suited for growing in other seasons.

For example, I've found that my trays of microgreens do very well in the kitchen during the autumn, when temperatures begin edging toward frost, but suffer in that room during the summer because the south-facing windows heat up the space too much. In those warmer months, the micros thrive in the basement, where I can control the light and air more easily, and avoid the humidity that makes growing more challenging.

Tomatoes and peppers, however, love the heat. Putting them by a large kitchen window in the middle of summer allows them to thrive, but placing them in a basement or cool attic space requires an exhausting amount of control measures to make sure they're happy. In other words, it's likely that your home has the right spot for whatever you have in mind; you just have to find out where that space might be.

*A quiet corner in a living room can be used as a garden or nursery. Here, seedlings are coaxed along under an LED light until they are strong enough to be transplanted into bigger containers and scattered throughout the house.*

*You wouldn't want to put a recliner or a desk in the small space between these doors, but a shallow shelf filled with garden plants fits the space very nicely.*

You can't do better than natural sunshine when it comes to providing a light source for starting and growing your edible indoor garden plants. But do be aware that too much sunlight will damage some more delicate plants.

# Light

For many indoor growers, some form of artificial light will come into play (more on that later), but for maximum efficiency and sustainability, utilize natural light as much as possible.

As a general rule, south-facing windows are preferred because they allow for abundant light, but depending on where you're located in the country, this could be a benefit or a drawback. Light streams in, but heat does, too. Placing some plants in direct sunlight during the hottest part of a summer day, especially without proper airflow, can cook them instead of bolstering growth.

When selecting a site for growing, look for one that allows for natural light, but can also be shaded in some way. This might be as easy as picking a window that has an awning outside that blocks the sun during the middle of the day, or placing plants on a shelf that gets indirect sunlight. Most likely, if you've noticed good results with existing houseplants, you've found some good spots already, but keep in mind that vegetables, herbs, and fruits need extra care like airflow and pest management.

Choosing a spot with natural light isn't mandatory, but it does cut down on the amount of work you'd have to put in for creating an all-artificial-light system. In my own grow space, I use sunlight as much as I can, by lining up plants on a window-level shelf in my south-facing kitchen and dining room, and then supplement with artificial lights in the winter.

# Airflow

Unlike many houseplants, indoor edibles need some type of airflow in order to grow properly. When I first started growing, I didn't realize the importance of this factor, and quickly saw the results of my knowledge gap: molding seeds, struggling starts, no germination, and bugs that seemed to come out of nowhere.

Air circulation helps to mimic outdoor conditions, helping plants to grow in a robust way while minimizing the risk of bacterial issues and pest problems. In my space, I'm fortunate enough to have a cross breeze from windows on two sides of my kitchen, but I still utilize small fans for days when there isn't much wind.

In areas like basements or attics, which can get stagnant pretty quickly, it's especially important to create better airflow. Check out the air circulation section later in this chapter for more in-depth strategies once you've chosen your primary growing space.

*A small desk fan can create enough ventilation for a few potted plants.*

# Room to Grow

Here's some bad news: you can't grow twenty different kinds of herbs on a three-foot space in a kitchen. Believe me, I wish someone had told me that a few years ago.

Like plants out in a field or in an outdoor raised bed, indoor plants need space apart from each other to stretch out. With the exception of microgreens and shoots, which are harvested during the first stage of growth and don't need ample room to expand, most indoor plants benefit from at least some breathing room. Herbs and many types of vegetables can be cozy, but they shouldn't be crowded.

When picking your growing spot and making a plan, create a rough sketch of where each pot or container will go, to give yourself a visual representation of your indoor garden. When it begins to feel like a game of Tetris, consider scaling back on the number of plants, in favor of giving the top contenders a better shot at growth.

If you plant your in-home garden near a water source, such as the kitchen sink, you'll have the option of bringing the plants to water. The main reason this is preferable is that it eliminates the risk of spilling water all around the plants set up throughout the home.

*While microgreens and shoots are planted very densely, herbs and veggies that are grown to full maturity require space between containers to allow for airflow and to give them access to direct and indirect light.*

To contain mess and water runoff, many indoor farmers move their plants to the sink area for watering and then return them.

## Water/Drainage

Thanks to the breadth of container types available, water and drainage isn't usually a major issue, but it should be considered with plants like microgreens, which have very specific watering needs. Locating those type of plants in a kitchen usually makes the most sense, since it's a few steps from a sink, where plants can be placed to drain or soak.

Another popular spot for just that reason is the bathroom, where the humidity levels can boost the health of some types of plants, usually those that grow in more humid zones. Dwarf citrus trees, for example, can thrive in a bathroom as long as there's enough air circulation and some natural light.

When considering the bathroom as a growing area, though, there are several notes of caution. One comes from my plumber, who's also an avid gardener: Don't soak plants in the bathtub of old houses, unless you want to be asked why you have vermiculite in your drains. Also, bathrooms tend to have the lowest level of light in the house, so some form of artificial light might be necessary.

Finally, think about all the product types you use in the bathrooms: spray-on deodorant, hair spray, perfume and cologne, talcum powder, and so on. Everything that goes airborne will affect the plants you keep in that room, and while that might be fine for houseplants, keep in mind that you'll be consuming the edible plants at some point.

## Humidity

Much like air circulation, humidity control is crucial when dealing with indoor growing. In some situations, like seed sprouting, abundant humidity can be very beneficial, but in others, such as pea shoot growing, it can make once-robust stems droop and fade.

Because of this, you may want to choose one spot in the house for seed starts—the bathroom, for example, or a sunny porch—and another for the majority of growing. This two-location approach allows you to develop a greenhouse space that can be better protected against pests, bacteria, and other issues that might affect tender seedlings. It's not always necessary to pick a greenhouse area that's hot and humid, but being able to control humidity in the area, either through plastic sheeting or individual domed lids for pots, can be helpful for making sure that plants get the best start possible.

## Pets & Pests

Sometimes, these can be the same thing. Although indoor systems benefit from being protected by outdoor critters like squirrels, rabbits, and chipmunks, one cat can become a mini-Godzilla to a burgeoning kitchen garden's Tokyo.

In our house, which is ruled unconditionally by two dogs, all we need to do is shut the basement door or move pots onto a counter instead of the floor. For other indoor growers with more wily pets like cats, the strategies may have to be more elaborate. I've seen a number of anti-kitty systems cobbled together by other indoor growers, and they can be impressive. Wire screens, large rocks, plastic mesh, repurposed bookshelves: suddenly, a

*A custom-made cage of poultry netting (otherwise known as chicken wire) can be fashioned around plants and containers to provide a layer of protection from curious pets. This may help, but on the downside, most pets are persistent enough to defeat this strategy and the cage definitely detracts from the loveliness of your indoor garden. Controlling access to the room or perhaps a few sessions at obedience school are better long-term solutions.*

once-simple indoor growth space looks like a kid's fort in the woods.

When choosing a space in the house, it's helpful to find a room that can be sectioned off easily, without scrap lumber becoming involved. This might be a kitchen where a swinging door is shut during the day, or a guest room that's already off-limits to pets.

In terms of true pests, this is far trickier. The aroma of fresh seeds can be compelling for critters like mice, and even houses that never had mouse problems before might be breached because of the new buffet you're creating. In the

*If it's green and growing, the chances are pretty good that cats and other pets will want a nibble or two.*

homes where I've lived, I've found that this is a problem mainly in unfinished basement spaces, and often during the colder months. Because of this issue, I don't grow in that type of area between September and April as a general rule. If that's the only space available, I mouse-proof all areas of the gardening area with as much creativity and kindness as I can muster. Fortunately, I haven't seen a problem in any upstairs spaces like a kitchen or dining room, perhaps because the

dogs are overly enthusiastic about seeing small creatures as new best friends.

There will be more, much more, about handling pests and insects in other parts of the book, but for now, it's best to try and choose a space that seems protected already. That can go a long way toward preventing anything other than you from eating your garden produce.

# Getting Started

Now that you've chosen a few prime spots in your house, it's time to create an optimal growing area that will be so amazing it will triple your dinner party invitee list, just so you can show it off in a faux-casual manner. "Oh, this space in the kitchen? Yes, I just thought I might grow some salad mix and herbs for tonight's dinner…you don't do that too?"

While practicing your humble-yet-talented gardener expression, here's what you should pick up to create a workable, efficient growing space that will lead to potluck ingredients galore: containers, soil, shelving, lights, fans/air, and of course, seeds or transplants. Let's break it all down.

*Even though you can repurpose about anything you can imagine as a planting container, there is something to be said for the utility of purchased items made for planting compared to an old rubber boot. Plastic plant trays with drainage panels are not expensive.*

*It may not be the ideal container for microgreens, but it sure is cute. The wisp of micros growing in this bottlecap demonstrate one thing quite clearly: you can grow microgreens in just about anything.*

## Containers

The breadth of container options is limitless; I've seen indoor vegetables grown in children's old sand buckets, retired purses, worn-out boots, tackle boxes, partially broken drawers, wicker baskets, even a rolling luggage bag that lost its top. Repurposing unused items that are just taking up space in the basement is a fun idea, and it works well for certain kind of projects, but there are a few items to keep in mind when digging through the junk room:

### Root depth

Every vegetable, fruit, or herb will have a certain root depth that it needs. Even though it's likely that you won't be attempting to grow deep-rooted plants like asparagus or artichokes (both of which have roots that extend at least 2 feet down), it's helpful to think about how much room you really need.

For microgreens, for example, harvest is done so early in the plant's growth stage that roots barely need any depth. In fact, some people grow them without soil at all, on algae-infused mats that provide nutrients and support growth up to a few inches in height. As a fun kid's project, you could grow microgreens in a bottlecap, or a teacup saucer, anything that can be gently watered and tended for about a week.

Most of the other candidates for indoor growing, though, do need some room for the roots to expand properly. For most vegetables, most of the root mass is within the top 6 inches of soil, so choose a container that can be aerated properly—no small-mouth jars, for instance, or other items that prompt soil compaction—and are reasonably sized for the project. You could choose a large, deep pot for small carrots, for example, but you'd end up using far more soil than you need.

So, choosing a container should have a sense of economy about it: Pick one that gives you room for root growth, but doesn't lead to soil waste.

## Drainage

Most likely, you already have plenty on hand if looking for repurposed options, but if you're shopping around at garden stores, consider containers that allow for drainage.

These containers have either a single hole in the bottom of the pot, or several small slits that stretch along the expanse of a tray or pot. Although I love the idea of repurposed containers, and I have a stack of potential candidates in my basement, I usually opt for rectangular trays with drainage slits.

Not only do these types of containers let me accidentally overwater without too much risk of mold, they also allow me to water a plant by setting the whole tray or pot into a sink filled with a few inches of water. For certain vegetables, especially if they're suffering from overwatering issues or leaf problems, "bottom watering" in this way is particularly helpful. Also, very dense plantings, such as a heavily planted pea shoot tray, benefit from this strategy. The roots can reach the moisture that they need without adverse effects on the upper plant.

There are some plants that don't need this kind of care (usually drought-resistant houseplants) but

*Commercial growing trays normally have drainage slots, which is important for growing indoors. But of course you must set the planting tray into a liner tray to keep the water from running everywhere.*

the majority of vegetables, shoots, herbs and other edibles benefit greatly from the aeration that comes as a result of drainage.

That's not to say that you can't take that antique lunchbox from middle school and turn it into a charming indoor gardening option—however, if plants seem to be struggling in that container due to root rot, then it may be time to go with better drainage strategies.

## Material

Many repurposed items are fine in terms of their material—old rubber boots keep the moisture in the soil, metal children's wagons make a nice mobile container, etc.—and many can be modified for drainage by drilling a few holes in the bottom.

One note of caution, though: keep in mind that you'll be eating whatever grows in these containers,

*Plastic and terra cotta pots are two of the most readily available types. Each has its pluses and minuses when it comes to growing edibles indoors.*

so if you're utilizing a cool, retro oil can, you'll have to be diligent in removing all traces of the oil before you plant. Even then, I would be reluctant to utilize that container unless it was my only option.

Any kind of container that seems suspect to me in terms of potentially toxic substances gets scratched off my list. It may seem like farm chic to plant in a rusty wheelbarrow, for instance, and it would be fine for non-edible plants, but I'm hesitant to expose the plant to rust, old chemicals, and other dangers because those issues may affect plant growth, but most importantly, those chemicals would be on my dinner plate, albeit in a small amount.

## CONTAINER TYPES

Some containers are more porous than others. Terra cotta pots, for example, draw moisture from the soil, so they require more frequent watering. Here are some other benefits and challenges of common container types:

*Plastic:* For most of my growing, I lean toward plastic even though it's not the most attractive or durable solution. These types of trays and pots are my preference because they're lightweight, stackable when not in use, generally quite cheap (or free, if you know a bunch of gardeners who are constantly trying to downsize), and easily modified with additional drainage holes. That said, they are still a petroleum-based product, so they're not the most environmentally friendly option available. I try to minimize my non-green impact by reusing them as much as possible, though.

*Polystyrene:* Although I don't use many white polystyrene foam boxes myself, I've seen numerous examples of successful growing with these. The trick is to find food-grade containers, but if you live in an area with any kind of supermarket diversity, it just takes a few phone calls to gather a nice collection. Most grocers, and all fishmongers, use these boxes but they have very limited re-use, so you'd be keeping them out of landfills. The foam provides excellent insulation, and it's easy to drill drainage holes in the bottom. The appearance is a drawback, but many people paint them so they look less like a Styrofoam cooler.

*Stone:* Most of these containers will likely be outside, since the largest drawback is weight. Smaller stone pots, though, can be a good option because they have a nice amount of heat insulation and are definitely very durable.

*Terra cotta/Clay:* In addition to being more porous, these pots can be prone to cracking when soil freezes. That's not a huge issue for indoor growing, but if you're storing them in a garage when they're not in use, be sure to empty the soil from them first. A more serious issue is that clay pots retain heat very effectively, which is great if you have plants that love heat (like peppers or eggplant), but for those that are less fond of long periods of heat, the plant roots can get burned. In general, though, these types of pots are visually appealing, fairly inexpensive, and have good drainage.

*Wood:* For many indoor growing spaces, wood is a nice choice because these containers can look amazing, especially with weathered wood (achieved by exposing it to actual weather, so throw a new containers outside for a few months). But wood can also be problematic when it comes to food safety—some containers are treated with very harsh chemicals to keep the wood from molding or rotting, and those toxins can leach into your edibles. Older containers in particular may have been treated with chemicals that leach arsenic into the soil. A good solution that I employ often is to put another pot inside the wooden planter, especially if the leaves will drape over the side to hide the gap between planter and pot. It's possible to seal the container with a good food-grade choice as well, such as a mineral oil and beeswax combination, or a soy sealer (brands include SoySeal or SoyGuard). Just be sure to avoid any conventional wood sealers, because they're chock full of chemicals that don't play well with edibles.

Basically, there's no single, perfect container type that works fantastically well for indoor vegetable and herb growing. Instead, choose a container mix based on what's safe for edible plants, attractive in your kitchen or other growing space, and easily watered.

*Many indoor growing media blends contain no soil or compost. A suitable media can be made from a blend of peat, coir, vermiculate, bark, and bonemeal.*

# Soil

Among the many variables that come with in-home growing, the importance of your soil mix can't be overemphasized. Seriously, your soil can make all the difference between healthy, vibrant, nutrient-dense vegetables and sickly, non-germinating, pale plants.

When I first started indoor growing, I thought: dirt is dirt, right? I can just go in my backyard, get a shovel full of the stuff, and I'm on my way. Forget the trip to the garden store and those fancy bags of soil—vegetables grow in dirt outside all the time, so obviously those soil mixes are a scam. This was followed about a week later by an exclamation of: hey, where did all these bugs come from? And about a month later by: hmm, I wonder why nothing is growing?

Very well-prepared soil that's geared toward indoor growing isn't a scam, and it's not optional. Homes and apartments tend to be drier than the outdoors, and putting plants in a container gives them less room and fewer nutrients than they'd have outside, even in a raised bed. Because of that, you need soil that's well aerated, yet able to retain enough moisture to foster growth.

If you use soil from the outdoors—even lovingly mixed with compost—drainage and aeration are almost always impacted, leading to poor growth, if you're lucky enough to get any growth at all. Often, garden soil is higher in nitrogen as well, which is fabulous for an outdoor garden, but when packed into a small space, it acts like a welding torch to your plant's roots.

Also, bugs. I think I could grow plants in a sealed room, ventilated with the purest filtered air, and handled only by volunteers in HAZMAT suits, and I would still get bugs if I used garden soil. That's because the insects are already in the dirt, so it doesn't matter what kind of pristine conditions you set up in your house, you're still giving them a free taxi ride to an all-day seedling buffet.

With those caveats, though, it's true that some people can pull it off. My partner, Karla, seems to be able to grow anything anywhere, and I believe she could nurse along a mango tree successfully through a Minnesota winter. She has the touch. But, because I don't, I tend to be more deliberate in how I set up my space for growing and with that in mind (in case you tend to be one of those people, too), here are some tips for choosing good soil:

- Some of the best options for indoor growing don't contain soil at all, but are instead a combination of materials like peat, bone meal, coir fiber (ground up coconut hulls), bark, and vermiculite.

- That last item can be particularly useful— vermiculite is a silicate that's fluffy and pebble-shaped. It helps to promote fast root growth, anchor young roots, boost moisture retention, and assist germination. You'd look for horticultural vermiculite, as opposed to other types that are used for shipping chemicals or enriching concrete. Many potting soils already have vermiculite added, however, and you can tell by the distinctive white, chalky flecks distributed throughout the mix. If you're whipping up your own mix, consider blending compost and vermiculite.

Choose a potting mix that is specifically for indoor vegetable and herb growing. These are put

*Vermiculite is a mica rock that is ground and then heated up until it explodes like popcorn. It has no nutritional value for plants, but it prevents the growing media from compacting.*

together in a way that fosters better drainage, and most of all, contains only a small amount of fertilizers. This is important, because too much fertilizer can burn a plant's roots, especially in smaller containers. I tend to use a compost-and-vermiculite blend for growing, and if it seems like any of my plants are struggling or slow in germinating, then I might turn to fertilizer like bone meal or fish meal, which are both great at maintaining proper soil chemistry without burning the young seedlings. Compost and fertilizer tends to work well together, because each supplies nutrients, especially during the early stages of plant growth.

This seems like a good point to climb atop my soapbox about organics. I love the view from up here. Simply put, we farm organically, and we believe deeply in creating a sustainable agriculture system, and that extends just as much to the cilantro growing on my kitchen counter as it does to the heirloom tomatoes out in our fields. So, I do my research on planting mixes, and I buy from locally owned greenhouses. Most of the time, this means that I pay more for my potting mix, but also that I feel better about what it contains.

I'm always keenly aware that what I grow will end up in my body, and a few extra dollars upfront is worth it to me to know that I'm getting as close to chemical-and-toxin-free as possible.

Just because something is labeled as "organic," though, doesn't make it natural and sustainable.

Recently, I went against my own eco-friendly policies because I was in a planting emergency (they really do happen!) and needed potting mix quickly. I opted for a stop at a big box retailer and got a suspiciously cheap mix that was labeled organic. Inside, I found a mix that was quite heavy and contained bits of candy wrappers, small wires, glass beads, and other weird garbage.

Since I was in a jam, I used it anyway, and I shouldn't have bothered. The mix was so dense that it turned my trays into mud blocks that wouldn't drain, and nothing germinated. When I dumped them out, the blocks had solidified into nearly unbreakable bricks of soil. Perhaps if I was building a mud house on the frontier, this would have been a valuable moment, but otherwise, it was a waste of time, money, and effort.

Choose a potting mix that is specifically for indoor vegetable and herb growing. These are put together in a way that fosters better drainage, and most of all, contains only a small amount of fertilizers. This is important, because too much fertilizer can burn a plant's roots, especially in smaller containers. I tend to use a compost-and-vermiculite blend for growing, and if it seems like any of my plants are struggling or slow in germinating, then I might turn to fertilizer like bone meal or fish meal, which are both great at maintaining proper soil chemistry without burning the young seedlings. Compost and fertilizer tends to work well together, because each supplies nutrients, especially during the early stages of plant growth.

# BOSSY E'S BEST INDOOR GROWING MEDIUM

Step One: measure out about 7 to 8 cups of indoor potting soil or compost mix, and about 2 cups of vermiculite.

Step Two: blend the dry ingredients thoroughly with your hands.

Step Three: add water; depending on how dry the soil is, you'll probably need at least 2 to 3 cups of water, but add it gradually and mix it in until you can squeeze a handful and have a few drops (not a stream) dribble out.

Step Four: test the consistency of the growing medium. It should clump together without too much pressure, but fall apart once you let go of it.

# Shelving

Of all the parts of a growing strategy, shelving usually has the most do-it-yourself appearance. I've seen "shelves" cobbled together from old screen doors, repurposed planks of wood on top of cinder blocks (making me recall my early days of college living), and bookshelves that have plants on top and slightly water-warped novels on the bottom.

Basically, it doesn't matter too much what you use, as long as it will support the weight of your plants and you don't mind getting it wet if you need to mist the plants or to put just-watered pots or trays on the shelves.

I've repurposed a wide range of shelving options, from countertop spice racks to old nightstands, and it's fun to root around in the basement and see unused furniture with new, savvy gardener vision. If putting together a system from scratch, though, here are some recommendations as a guide:

- *Consider metal, industrial-style shelving.* This type of unit is usually sturdy, able to be configured easily, and best of all, has a nice degree of openness in the back and sides. Also, steel shelves have an open slat design for each shelf, which helps to increase airflow. Especially handy, these open slats let you hang lights fairly quickly without the kind of drilling or clever DIY effort you'd need for standard wooden bookshelves.

- *Don't put plants on shelves that you love.* This sounds like odd advice, but believe me, any surface will get dirty and wet very quickly. Even when you think you can be extremely careful in keeping the bottom of the pots and trays dry, you can still get water stains on wood shelves.

- *Think about your lighting options when buying shelves.* We'll dive into lights in the next section, but for now, all you have to remember is that a standard-sized shop light is 4 feet

*You don't have to make your living room or kitchen feel like a commercial greenhouse to grow edibles indoors. Tasteful wall shelving that's integrated into your room décor does the job, too.*

*Take the edible landscape idea indoors by combining your edibles with houseplants and even cut flowers in your existing shelving and display furnishings.*

*Coated wire shelving is inexpensive, durable and adjustable. It also allows good airflow.*

long. If you buy an industrial-type shelving unit that's 48 inches across, and has the open slat design for hanging those shop lights, you'll feel like a pro. At least, that's how I felt after years of weird systems that I patched together, many of them featuring shop lights hanging way past a shelf's edge.

- *Go for adjustable shelving.* Being able to change the height of your shelves is very useful, because plants need different levels of light. It's possible (and recommended) to use the chains that come with shop lights to adjust height of lights, but it's also a nice option if you can re-configure shelves as well. This comes in handy, too, when using one of the shelves as a storage area for tools, extra pots, a watering can, and other growing supplies. Quick note on storage: if you'll be keeping your seeds on this shelf, make sure to store them in a sealable plastic bin, to lessen the temptation to pests.

In general, find a shelving system that allows for a nice amount of light (natural or artificial), gives the plants room to spread out a little, and offers enough space for proper airflow around the plant. As mentioned previously, lack of airflow is one of the biggest causes of nasty issues like mold and disease, so keep it in mind as you're creating your shelving setup.

Also, if you're just starting out, be aware that you may have to move your shelves if your chosen space turns out to be less than ideal. A spot that seems like it would be perfect may end up being too close to the window, or not situated near an electrical outlet for your shop lights, or too cold and damp. Being flexible about placement (here's where the wheeled shelf comes in handy) is all part of the adventure when it comes to indoor growing.

# Lights

When I teach classes about growing vegetables, microgreens, and herbs indoors, the question of lighting always comes up first, no matter how early I put it into my talk. At this point, I should probably just issue a note of assurance before the introduction: "I promise, you don't need special, expensive grow lights that are hard to find and burn out easily. Really. I promise."

In fact, all you need are full-spectrum fluorescent bulbs, similar to what you see in office buildings. They can be put into standard metal light fixtures that have one bulb in them, and feature small chains on either end for easy placement into a shelving setup.

These are found in any hardware store and are called various names like utility lighting or shop lighting. Occasionally, they go on sale, and this is when you should pounce on them and buy several at a time if you're going to create a shelving unit filled with plants. Otherwise, you can probably get one, with bulb, for under $30 and usually for much less. Thanks to buying bulbs in a bulk contractor pack (also at the hardware store) and a sale on the fixtures, I paid about $10 for each of my lights. That's quite a difference from the LED grow light systems, where a single bulb can be $60.

You may already have proper lighting if you're planning on growing vegetables, herbs, and fruits in your kitchen. Most under-the-counter fluorescent lighting works well for fostering growth, and all you'd need is some type of small block or shelf that brings the plants closer to that light. I have this type of lighting in my kitchen, and I like to grow herbs in smaller pots so that I can use a little shelf that used to hold dried spices.

*A hanging trouble light with an incandescent bulb emits more heat than light, which isn't necessarily a bad thing for growing plants.*

Beyond that basic setup, there are other lighting options if you have specific growing needs—sometimes, plants need a boost in some way, and that's when you can try playing around with lighting options.

For example, if a plant seems to be spindly or too leggy, you can give it a burst of red or orange light, which stimulates growth and flowering. If the plants are too short or stocky, you can use blue or green light to regulate plant growth or foliage.

Here's a quick guide to your options:

## Incandescents

- Used most for low-light, heat-loving house-plants like ferns or vines

Only about 10 percent of their energy used for light, with the rest used for heat, so they tend to be inappropriate for indoor edibles since they "cook" the plant.

## Full-Spectrum Fluorescents

- Best replicates the natural solar spectrum
- Life span of 24,000 hours
- Good choice for year-round growing
- When buying these, look for T8 or T5 bulbs

*Full-spectrum fluorescent lighting most often is found with thinner T8 style bulbs like the one above.*

## Fluorescents

- Two to three times more light than incandescent bulbs for the same amount of energy

- Least expensive option

- Tends to be the most available, especially if you're considering using existing under-the-counter kitchen lighting or shop lights

- Life span of 16,000 to 20,000 hours

- When buying these, most standard is a T12 bulb

*Fluorescent tubes are efficient and available in full-spectrum light models that are intended to replicate natural light as closely as possible. The standard T12 style bulb, like the one above, is by far the most common, but this style is gradually being phased out in favor of more efficient, narrower tubes that operate with an electronic ballast. The T12 works with a magnetic ballast.*

*A plug-in shop light with a full-spectrum fluorescent bulb makes a perfect light source for indoor gardening. Look for one that has hanging chains for easy lowering and raising.*

*LEFT: HID bulbs fit into standard sockets, such as this hanging trouble light. Because the bulbs are filled with an inert gas they operate very efficiently.*

*RIGHT: LED lights are extremely energy efficient and come in an ever-expanding range of sizes and configurations. While any light will do to some extent, you'll get best results from an LED grow light that is designed to emit the complete spectrum of light wavelengths that plants require.*

## High Intensity Discharge (HID)

- More expensive than fluorescents, but twice as efficient

- Can get bulbs that are only red/orange (called MH bulbs) or only blue/green (called HPS bulbs) so you can tweak your growing conditions

- Color of the lights distorts the appearance of plants and grow space; a minor consideration, but some gardeners are put off by the tendency of HID lights to make everyone look jaundiced

## Light-Emitting Diode (LED)

- More expensive than fluorescents

- Some bulbs don't require special fixtures, since they're already placed inside panels that can be plugged in.

- Tend to be more appropriate for commercial use

**TURN IT OFF**

Nerdy gardening facts: plants measure the duration of light and dark through a pigment called phytochrome, and their response to the relative length of a day is called photoperiodism. This response is vitally important for growth, seed germination, and vegetable or fruit formation. So, no matter what type of light source you're using, be sure to switch it off at night so the plants get their rest.

The newest grow light technology is plasma, which purports to be as close to real sunlight as possible. However, I haven't seen one of these systems for under $1,500, so for now, if I want my plants to have more sunlight, I'll just shove them closer to the window. In general, I prefer the full-spectrum fluorescents because they're affordable and available, but if you want to try an HID or an LED system, then by all means, go for it.

One important note when it comes to lighting: just like people, plants need a night as well as a day. Some growers have tried to boost growth by keeping the lights on at all times, but I've found this creates unnecessary stress on the plants because they don't have time to "sleep." You wouldn't water them with Red Bull, so why would you keep them sleep deprived? A better strategy, I've found, is to turn the lights on in the morning and turn them off about the same time as sunset; in the winter, I'll keep the lights on longer in the evening, but I'll be sure to turn them off before I go to bed.

## Fans/Air

When crops are outside, they get a little blown around by the wind, and this is crucial for more than taking pretty farm pictures of swaying corn and sunflowers. That airflow assists with temperature control, humidity, disease resistance, and oxygen intake. Plants grow stronger from being "stressed" by the wind, and they end up more robust as a result.

Growing plants indoors, you need to find a way to replicate natural conditions—similar to setting up an artificial day for your vegetables and herbs, you also need to create an artificial wind. Fortunately, you can do it on the cheap.

If your plants are near a window and the temperature outside isn't extremely hot or bitingly cold, simply opening the window for a few hours per day can yield enormous benefits.

When plants are in a more enclosed area without a window, it can be trickier to adjust airflow, but certainly not impossible. For vegetables grown in our basement, for example, I set up two box fans and arranged them so the airstreams would cross, set on low. This made the seedlings waver in the "wind," but not get mowed down by the strength of the breeze. The circulation is crucial, because plants don't do well if air is directed in a solid stream in one direction. That would be like taking a hairdryer, putting it on the coolest setting, and aiming it at your seedlings.

*For vegetables grown in our basement I set up two box fans and arranged them so the airstreams would cross, set on low.*

*A small desk fan, like this retro-style, oscillating model, is sufficient to provide a light breeze to help keep a small grouping of edibles healthy. Generally, the more your growing conditions mimic the outdoors, the better your plants will do.*

In another room, where there was a window on either side but I didn't put the plants close enough to benefit from opening them, I put a fan in each window—one angled toward outside, and one angled into the room. This sets up a simple intake and exhaust system, where fresh air came in regularly and stale air vented out the other side. In order to break up the airflow and create cross-ventilation, I set up a small oscillating desk fan above the plants, at the same level as the intake fan.

If you already have plants in place and are starting to notice any fungal issues or disease problems (like mold), and you've worked to make sure you're not overwatering, chances are good that airflow will be able to help. Buy a few cheap oscillating fans or box fans and set them up so that you can get some ventilation in your growing area during the day, about the same length of time as you have the lights on. If the air seems very stagnant, though—which can happen in an area like a basement—it's fine to leave them on at all times.

In general, the idea is to mimic the low levels of wind that a plant would get while outside. This can help reduce incidence of disease, and make vegetables and herbs heartier.

# Seeds

Every winter, it seems that the seed catalogs arrive at roughly the same time, leading to a pile of reading material that always makes me geek out with happiness. I pore over each description, look longingly at the heirloom varieties, and make my (often overly ambitious) plans for the summer. During a part of the year when the only sound seems to be wind and snowplows, the seed catalogs provide a delicious glimpse of abundance.

They're also hugely helpful for indoor growing as well, because beyond the luscious descriptions, these catalogs—both in print and online—provide a wealth of information that can be used to plan a better indoor garden strategy.

A quick note on organics here: because we own a farm that's been certified organic, we must buy organic seed, since that's part of the certification process. Beyond that, I'm a fan of organic growing, so even if we didn't have to use those type of seeds for Bossy Acres, I'd still buy them. I believe that organic practices lead to more sustainability, healthier soils, and a better agricultural system in general. Because of that, I think organic seeds are worth the extra cost that's usually involved in purchasing the seeds. You might opt for inexpensive, non-organic seeds instead, and that's cool, no judgment.

I'd just advise you to make sure your seeds are coming from an established source, where you can get the type of growing information that you need. There have been many times that helpful family members have given me seeds from who-knows-where and they hand them over in little plastic bags. "These are hot peppers," one of them will say. "They're really good. I think. They might

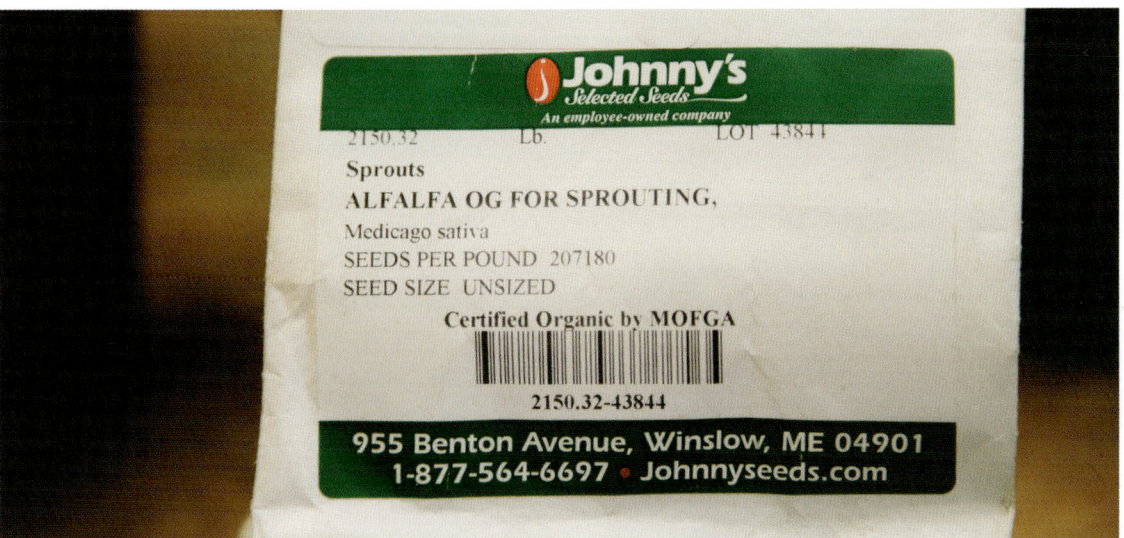

*Just about any of the seeds you'll use indoors are available as organic or conventional (non-organic). Both work, and while conventional are cheaper and easier to find, choosing organic seeds is a great way to support organic growing practices by voting with your wallet.*

be sweet peppers, though." That's the extent of the information I receive—they don't know the variety, growing timeframe, potential root depth, or anything else that helps me make a decision in how I grow the plant. So, I usually just end up putting them in my backyard's raised beds, in a grab-bag experimental area that doesn't get much devotion or tending. Sometimes it works out, most times it doesn't.

Another consideration in choosing seeds is whether to opt for heirloom varieties, even though some growers have found that heirlooms may be less consistent in terms of yield. Still, they're fun for me, because it feels like I'm continuing a tradition and keeping plant variety going. Heirlooms are plants that aren't grown in commercial-scale agriculture, and hail from previous generations. For example, there are only a few potato varieties grown on large-scale farms, which makes it enjoyable for me to choose lesser-grown varieties that might "die out" if it weren't for dedicated growers who keep them going.

Keep in mind that not all heirloom seeds are organic, but if you want both heirloom and organic, there are many options. Take a look in the Resources section of the book for a short list of seed providers that I've found dependable. Basically, seeds are everywhere if you start looking for them; once, I saw a seed rack in the gas station of a small town.

No matter where you obtain your seeds, be sure to store them properly, in a plastic bin with a secure top. This will prevent numerous pest issues, and help to prolong the life of your seeds. For best results, use the seeds within a year or so, and sooner if possible. The older seeds get, the less likely they are to germinate. Sometimes, when I've had seeds around for a while, I create an "anything goes" mi-

*A small plastic file box makes a great storage container for seed packets. Keep the box in a cool, dry area away from direct sunlight.*

crogreens mix, since I'll harvest at the first stage of growth anyway. If I want actual vegetables, though, I go with fresh seeds whenever possible.

Another handy tip: jot down notes right on the seed packet, including when the seeds were purchased. I use the packets for observations as well, noting what might be fast growing or whether a variety proved to be particularly good for one of my vegetable fermentation projects. It's very easy for me to lose track of notebooks, even when I try to store them with my growing supplies. But because I keep my empty seed packets in the same bin as my other seeds, I know where they are, and that one small packet will be rife with information, from both the seed company and my own experience.

# IMPORTANT INFORMATION ON BUYING SEEDS

This is what you need to know when choosing seed varieties:

*Light requirements:* Similar to houseplants, it's useful to know the level of light an edible might require. For example, beets do well with full sun every day, but chard can thrive with some occasional light shade.

*Seeding depth:* This is usually more important when you're seeding outdoors, but is still worth knowing if you're planting inside so that you don't push the seed too far into the pot and lessen your chances of germination.

*Plant spacing:* Again, this information is better suited to row gardening or farming, but if you're putting multiple edibles in one pot (such as lettuces or herbs), it's good to know so that you don't crowd out the plants before they've had a chance to establish.

*Days to maturity:* This is one of the most crucial pieces of info for me, because it can help me gauge when I can expect to harvest. Even with plants that regenerate, like herbs, I like to have a sense of the timeframe from seeding to harvest.

*Soil and fertilization needs:* Many seed companies will provide acidity level information, which is great if you love to dig into the science

of growing, but it's not mandatory to check those levels in order to have a good growing strategy. However, do pay attention to what type of soil works best—loose, well-drained soils are usually recommended for container gardening—and whether you need to use any type of fertilizer. Since nutrient requirements vary widely according to plants, we'll delve into fertilizer needs later in the book when discussing specific crop varieties.

*Plant management:* Some seed companies do a fantastic job of providing tips on pests, diseases, harvesting, and even storage. Reading through these descriptions can feel like a college agriculture course sometimes, and I've walked away from a seed catalog reading session knowing about things like pirate bugs (not as adorable as they sound).

*Container gardening suitable:* Because of the rising interest in indoor growing and container gardening outside, seed companies have started putting some great information on their websites. High Mowing Organic Seeds, for instance, has a nice online section about the topic, including suggested varieties, tips on growing, and a "seed collection" of 10 packets that tend to do well in containers.

*Transplanting a new seedling from a nursery can be done simply by setting it into your container and filling around the roots with growing medium. If you are transplanting a small plant or seedling that you've grown outdoors, remove as much soil as you can from the root system to minimize the chance of introducing disease.*

# Transplants

Although there's a certain thrill with growing from seed—seeing the germination and that first little pop of a plant start never gets old for me—transplants might be more useful for some indoor growing. Particularly if you've had difficulty starting from seed before, or you just want a jumpstart on your growing plan, transplants from a reliable source are a nice way to build your garden quickly.

In some cases, especially for herbs, it makes sense to transfer the plants from an outdoor garden into your indoor growing mix. If you have some robust rosemary outside, for example, there's no need to start rosemary inside from seed since you can just take a cutting off that plant or dig it up completely and plunk it into an indoor pot.

When bringing a plant from outside to inside, make sure to knock as much garden soil off the roots as possible, without damaging the plant. Even if the plant has thrived in this soil for months, there's risk of transferring insects or disease that can affect other parts of your indoor garden. Also, outdoor soil doesn't drain well when used in pots or planters, so use a potting mix instead.

Another consideration is the transition temperature. Whenever you move a plant from an indoor environment to an outdoor space, you need to make that shift gradual so that the plant has time to adjust, and the same consideration is helpful for bringing the plant in the other direction. Re-pot outside, if possible, or in a cooler space like a garage. Water well, and then plan on leaving the plant in a transitional area for a few

*Timers are plugged in to your receptacles and programmed to turn lights on and off (you plug the light into an outlet on the timer body).*

*Heat mats are placed beneath planting trays to warm the plant roots and to speed up germination in cooler weather.*

days—garage, porch, enclosed deck, anyplace where the temperature is slightly warmer than outside. If available, make one more transfer before bringing the plant inside.

For example, if I were to re-pot an herb from outside, I'd put it in the garage first with the windows open for better airflow, then after a few days I'd transfer the plant to my enclosed porch, near one of the open windows in there. Finally, I'd bring the plant inside after a week to 10 days, making sure to water thoroughly in order to aid the transition. Like many people, plants don't respond well to sudden changes, so taking a gentle approach is always helpful.

If using transplants from a garden store or farmers' market, then it's fine to bring them directly inside. The plants have already had a chance to be in an environment that's cozier than outside in the ground, and they're prepared for indoor growing.

Whether your transplants come from a store or your own outdoor space, also be sure to pot them up in a container that's free of disease. Using a new pot or tray works for this, but you can utilize what you have on hand as long as you make sure that the container's previous tenant didn't have any issues. If you're throwing out a plant because it never thrives, for example, or seems to always have disease problems, it's best to recycle that container as well—even if the plant is gone, the disease can linger, so it's better (and more economical) to be safe.

*Mister bottles come in many forms, from very cheap to professional grade. It's worth investing in a better quality model.*

# Less Common Problems

## Miscellaneous Growing Supplies

Even when you're working to keep costs in line, it seems there's always just one more thing to buy at the garden center. Maybe you want to try some coconut coir to mix into your soil to improve drainage, or you're loving those glass watering bulbs that can be stuck into a pot and left for days. When I'm trying to stick to a budget, I try to avoid going to the garden center for anything because I feel like one of the kids who saw Willy Wonka's candy garden for the first time.

In order to properly outfit your little growing space, though, you do need a few supplies, and these come in handy:

- *Automatic light timers.* If you don't want to worry about turning lights on and off during a specific timeframe, these timers are very useful, and they're often fairly inexpensive. I use them when I'm going away for a weekend, or if I'll be busy for a stretch of time and want to reduce my indoor garden maintenance.

- *Heat mats.* Also called germination mats, these are designed to be placed underneath plants so that roots stay warm in cooler areas like basements or drafty corners. They're usually rectangular, and don't have temperature settings; you just roll them out, plug them in, put your pots or trays on top and you're good to go. The low temperature won't burn your plants (unlike, say, a heating pad

would) and they do help during the winter months, I've found. Also, if your plants seem slow to germinate, using one of these can provide a boost. One caveat: they're usually not cheap, so if you see one on sale, grab it.

- *Plastic mister bottle.* These are extremely useful, and only cost a few dollars. They can be utilized for an array of growing-related tasks, like spritzing fish emulsion on tomatoes, or spraying a soapy solution on

*Living indoors does limit the type and number of afflictions that can affect plants, but they are by no means immune from problems such as mold and disease. In most cases, the remedy is more or less water and/or light.*

aphid-ravaged plants. During certain times of year, my growing space feels a little drier than other times, so I often mist water over all the plants at least once per day.

- *Small plastic bins/totes.* Usually, I can get these for just a few bucks from places like Target, Family Dollar, or IKEA. Whenever there's a sale, I must look like a professional organizer, because I load up on them, especially ones that are about the size of a shoebox. Put everything in these that you need: scissors, twine, Sharpies, pens, light timers, rubber bands, etc. Also, these bins will be handy for organizing seeds; I tend to use several containers so I can sort seeds according to usage (pea shoot seeds in one bin, micro-greens seed in another, and so on). I really wish that the rest of my house was as organized as my indoor gardening space.

In the rest of this book, I'll dive into details on how to grow specific types of indoor vegetables, herbs, and fruits, but there are common issues that affect most indoor edibles, and it's helpful to know these general concerns before planting that first seed.

## Mold

Vegetables, herbs, fruits, and even houseplants are all more susceptible to mold issues than outdoor plants for a number of reasons. They tend to get more water, which can introduce mold spores, and they're sometimes growing in wooden containers, which introduce that water into small cracks in the wood. Humidity, airflow, and poor soil also contribute to mold.

It's quite easy to tell that you have a mold problem, because if you've ever seen a loaf of bread

One homeopathic remedy for abating mold on your indoor plants is to crush a clove of garlic into a cup of water. Let the garlic steep for 15 minutes or so, then transfer the garlic-infused water to a mister and spray the affected plant lightly.

go bad then you know what can happen—a gray or yellowish fuzz begins forming in just one spot and soon it's expanding everywhere. Getting mold on the plant itself is less common, fortunately, but it does happen if mold on the soil isn't treated in a timely way. Most often, you'll see that telltale fuzz at the base of the plant, or along the container edge.

Prevent mold by watering when the plant needs it, not as a daily (or twice daily) habit. There are many conditions, such as cooler temperatures, when plants don't need as much moisture, so putting plants on an automatic watering schedule can result in dampness, leading to mold. Another problem is inadequate drainage. When this is lacking, plants are sitting in soggy soil, which can lead not just to mold but also insect issues and disease.

Putting the plant in more direct sunlight can be helpful, as well as increasing the airflow or putting the plant near an open window so it can get more fresh air.

There are also chemical fungicides, but I like to consider those as a very last resort since I'm spraying them into my indoor air, even if it's a couple spritzes. Instead, I've had some success with crushing up a clove of garlic and letting it sit in a cup of water for about 15 minutes; then, I put the concoction in a mister, shake thoroughly, and spray on the plants. That method, combined with moving the plant to a sunny spot, often does the trick.

Also, some essential oils can be used as natural mold fighters, including rosemary, cinnamon, and tea tree. Just put a few drops into a cup of water, add to mister, and spray the affected areas—for stubborn mold, you can spray once a week.

No matter what the method used, just be sure to isolate the plant so that the mold doesn't spread to other vegetables, herbs, and fruits.

If mold keeps cropping up—and for some people, it's a scourge that seems to plague their indoor growing efforts on a continual basis—consider switching out your setup completely, especially your soil and containers. It can be pricey to do a total rehaul of your efforts, but a fresh start can help to tackle ongoing mold issues.

## Pest Control

As a farmer, and especially as an organic farmer, I'm used to seeing an array of bugs, from flea beetles that feast on my cabbages to Colorado potato beetles that can strip a plant down to sticks in less than a day. Even with the knowledge that we'll be losing part of our crops to insect damage, however, I still feel an elevated sense of injustice about them, as if they're the mean girls in my otherwise peaceful high school lunchroom.

That same smoldering frustration comes up in my indoor growing efforts, but at least my at-home gardening is on a much smaller scale, and so I can actually do something about the situation. The usual suspects for indoor pests include aphids, whiteflies, mites, and mealybugs, all terrifically annoying.

No matter what options you choose, it's smart to gravitate toward non-toxic options. Although I'm obviously a fan of organic methods, this goes beyond my beliefs—conventional pesticide sprays that are designed to be used outside (and usually with respiratory protection) can wreak havoc when used inside where air ventilation is severely reduced.

Also, much like choosing a non-toxic container for planting, keep in mind that whatever comes in contact with the plant may eventually affect the person or animal who eats it. Personally, I don't like the idea of nibbling on chemical-laden pesticides, even in trace amounts. So, I choose not to go that route. But if you're finding that nothing works whatsoever and you want to try to save the plant from infestation, then at least bring the plant outside if you spray it with conventional pesticide formulas, so that the toxins don't get trapped inside the house.

### SAFETY NOTICE

Never use pesticides intended for outdoor usage indoors.

## PREVENTING DISEASE

Here are a few pest control measures to try on any kind of vegetable, herb, or fruit plant:

Increase the air circulation by placing a fan nearby or opening multiple windows that allow strong cross breezes.

Move adjacent plants, especially houseplants, away from your vegetables and herbs—the pests may be targeting one type of plant but end up affecting your whole indoor garden.

Dip a cotton swab in rubbing alcohol and gently brush the plant's leaves, if the insects are mainly on the top of the plant and aren't too numerous.

Create a mix of mild liquid soap (like Dr. Bronner's) and water and spray the plants.

If the outdoor weather is amiable, put the plants outside; if you're dealing with aphids, you may have a hungry ladybug population outside just waiting for this kind of takeout.

It's possible to buy beneficial insects like ladybugs or lacewings and release them inside your house without causing a rippling effect (in other words, you're not getting a mongoose to handle the snake that you bought to kill the mice), but keep in mind that this method tends to work best with large-scale pest populations. If you have a sizeable greenhouse area, for example, then it might make sense, but if you've only got a few aphid-infected indoor tomato plants next to your kitchen window, buying a standard order of ladybugs (at least 4,000) seems like overkill, to put it mildly.

*To discourage insects from munching on plant leaves, mix a small amount of mild liquid soap with warm water and lightly spray the plant with a mister. This won't harm the plant but makes the leaves quite distasteful (be sure to wash them before eating them yourselves).*

On a preventative note, carefully inspect any transplants that you're planning to bring inside for indoor growing. If leaves are shiny, or tiny eggs appear in clusters anywhere on the plant, clean it off carefully and leave it outside for a few days if possible to make sure the pests don't hatch into a full infestation.

In terms of other pests, mice can be an issue, too. They tend to appear in older houses, I've found, as well as greenhouse spaces that are attached to kitchens, because those rooms may have air ducts close to the ground. Nearly every greenhouse I've ever visited has mousetraps tucked into the corners, and birds swooping in and out through open windows to steal seeds.

Controlling these pests is very difficult, but not impossible. Some anecdotal evidence suggests that sprinkling peppermint oil around pots and trays is effective, as well as a box-type device called "the tin cat," in which mice can enter but not exit. The ultrasonic devices, however, have been universally panned by nearly everyone I've asked (and believe me, I've had way too many mice discussions in the past few years), and most people just suggest I get a real cat.

# Disease Issues

As disheartening as mold or mice are shriveled or mottled leaves, blackened vegetables, or other indications of disease can also be dismaying. Unfortunately, disease can occur even in the best setup, and greenhouse managers are often on a constant state of high alert to try to prevent issues. At our farm, we make sure to wear greenhouse-only boots, and to rinse off any tools or carts that have been in the fields, to prevent diseases from hopping inside from outside.

In a house or apartment, measures like that aren't feasible. You may be transferring disease because you didn't shake off enough of the outdoor soil when repotting herbs from the garden, or your plants might be more susceptible to issues because your care isn't quite what's needed.

*Powdery mildew is one of the more common fungal infections plants experience in agriculture, but it can hit your indoor garden too. Overly humid conditions are a prime contributor to its growth. One organic treatment that growers suggest is spraying the plant leaves with a solution of 1 part milk to 10 parts water at the first signs of mildew.*

# Common Disease Problems

When it comes to indoor growing, there are several types of diseases that can come into play, and here are general disease categories worth noting:

**Fungal:** Simply put, these involve a type of fungus that's trying to take over your plant's leaves and stems. Most often, this results in black spots, grayish mold or fuzz, rot at the base of the stems, or whitish powder coating leaves and stems. Sometimes, the plant seems fine and then suddenly wilts and dies within a short timeframe, which is an indication of root rot. Another fungal problem is "rust," which manifests as small, rust-colored bumps on the underside of leaves that eventually turn yellow.

**Bacterial:** When a detrimental bacteria affects your plant, the decline can happen fairly rapidly as the plant tries responding to the invader. The most common symptom of a bacterial problem is leaf spot, which can create round areas of dead cells on foliage, indicating infection. You might also see odd-looking bumps on stems, or abnormal growth on leaves and roots.

**Viral:** If a plant is suffering from a viral issue, it's most likely that you have an insect problem, since that's the most common way that plants get infected. Bugs feast on a diseased plant and then jump over to a healthy one and spread the disease when they begin their new meal. But viruses can also be spread by people, or even come from infected seeds. Plants with viral problems often appear stunted in some way, or misshapen. Leaves might roll up, or appear yellowish or white.

*Fungal infections that select plants quite specifically are fairly rare in indoor gardens. The Cercospora beticola infection visible on these beet leaves is commonly known as "sugarbeet leaf spot" and can be combated only with harsh chemical pesticides that should not be used indoors. In most cases, the best action if you discover disease in your indoor garden is to discard the entire plant (and the soil it is in) before other plants are affected.*

**Abiotic:** This is the term used when environmental factors cause problems rather than organisms like bacteria, fungus, or insects. Abiotic issues range from salt burn—caused by too much fertilizer in the soil—to sunburn from excessive light. I've sometimes seen frost damage occurring when a plant is put too close to a window during cold weather; the leaves might touch the chilled glass and then turn black as a result.

# Prevention and Treatment

Reading about plant diseases can begin to feel disheartening, like browsing WebMD when you have a cold and walking away convinced you have scurvy. But don't worry—there are ways to prevent issues, and deal with them if they come up:

- Don't overwater: In the same way that mold can form from overwatering, diseases can find a hospitable environment in an overwatered plant.

- Don't overfertilize: Although you may think that more nutrients are better than fewer, there's danger in putting too much fertilizer into your plants. Specifically, the excess nitrogen can weaken a plant and affect its ability to resist diseases.

- Ventilate: I know it seems like I keep harping on airflow, but at the risk of constantly repeating myself, think about airflow. Ventilation is your friend.

- Space properly: When you crowd plants too close together (except for microgreens and shoots), problems can occur, so be sure to adhere to the spacing requirements for each plant based on seed company information or advice on specific crops found later in the book.

- Isolate: When dealing with a disease in a plant, move it away from your others, even if they're different types of plants. This helps to keep the disease contained to one area, and makes it easier to treat just one plant instead of several.

- Throw out: Sometimes, you put a great deal of effort into a project and it's just not working. If disease is the culprit, it's better to discard the plant and start fresh, because you don't want your other plants to be affected. It's also advisable to use the pot or tray for other purposes (pencil holder, perhaps, or an in-box for mail) because even if you thoroughly wash it, there's always a slight possibility that you didn't get all of it. I find it's better to be paranoid than to go through multiple rounds of fighting the same problem.

# PART TWO

# Quick and Easy Greens

There are many types of edibles that are traditionally grown outdoors, but can be planted inside for fun—more on those in Part Three of the book—but there are also several kinds of "crops" that are ideally suited to an indoor growing environment. Specifically, microgreens, pea shoots, sunflower shoots, sprouted grains, and some herbs can thrive inside much better than they could fare in a garden.

These plants are often harvested at an early stage of a lifecycle, and they tend to be more fragile and delicate than what's growing outside in the garden. Sprouts, in particular, might do very poorly in outdoor conditions, even on a patio or deck, where sunlight early in the growing process might hinder germination.

For those who are new to indoor growing, the projects in this section are ideal as a starting point, and the quick seed-to-harvest timeframes mean that you won't be spending months waiting for vegetables to emerge. In fact, some microgreens can be harvested in less than a week, if factors are right.

These projects also help to determine if your indoor space is well suited for growing. Because of short timeframes, you'll be able to spot problems like airflow and overwatering more easily. For example, too much watering and stagnant air can create a mold situation in sunflower shoots or microgreens within just a few days, prompting the need to tweak your setup.

As with any indoor growing efforts, enjoy the process—especially when those pea shoots or fresh herbs can be thrown into dishes just moments after you harvest them.

# Microgreens

Although some seed companies offer mixes designated as microgreens, there's no such thing as a "microgreen seed." They aren't grown using some special, almost magical seed that will grow a plant that's only about three inches in height. Instead, microgreens can be grown from nearly any seed, since they represent the first stage of growth of a plant.

Called cotyledons, these initial leaves of a seedling give way eventually to a plant's "true leaves," and from there the growth truly begins into vegetable, herb, or fruit. In other words, if you plant seeds in order to get microgreens and then change your mind or leave them for longer than intended, the plant will begin maturing and, most likely, get too large for the pot you've chosen. Most micros, by loose definition, are the cotyledon and first true leaves of a plant, although they can also be harvested at only the cotyledon stage.

Microgreens look similar to sprouted greens—and some online growing resources put them together in the same category—but unlike sprouts, which are grown in water, microgreens are grown in soil. They do best indoors, since trying to adjust planting density and moisture in outdoor conditions can be tricky.

In terms of arrival on the home garden scene, microgreens are the new kids on the block, but their popularity with chefs, small-scale farmers, and urban growers will likely propel microgreens past the trendy stage. Their cute size, mule-kick-level flavor, and nutritional clout make them a perfect addition to any indoor growing mix.

*When used as a garnish, microgreens add intense flavor and dense nutrients to savory dishes. This fried pork dish gets a big boost from a sprinkle of mustard microgreens.*

*Talk about instant gratification. Most microgreens, including these peashoots, are ready to eat less than weeks after planting.*

Although they're appearing more often on the menus at upscale restaurants, microgeens only started becoming popular in the last few years, and even the word itself is fresh. Mark Mathew Braunstein, author of *Sprout Garden* and *Radical Vegetarianism*, has noted that first use of the term "microgreens" was in 1998. By comparison, the first known use of the word "radish" appeared in the fifteenth century.

There's a reason microgreens are catching on quickly. According to a recent study in the *Journal of Agricultural and Food Chemistry*, microgreens can have up to forty times more nutrients than mature plants, ounce for ounce. Their nutritional density depends on the type of microgreen, but all seem lush with nutrients. For example, red cabbage microgreens showed very high levels of vitamin C, vitamin K, and vitamin E. Considering how easy it would be to slip some micros into a kid's meal for an extra nutrient boost, keeping a small tray going in a kitchen garden makes sense.

Most of all, though, there's the flavor. A single, slender beet microgreen only as long as a fingertip, for example, can taste like a fully grown beet. Mustard and radish microgreens are far spicier than many people might expect, and carrot microgreens carry the fresh, sweet flavor of just-harvested vegetables. They sometimes seem like the gum stolen by Violet in *Charlie and the Chocolate Factory*, when she tastes an entire dinner in a single stick. (Fortunately, unlike Violet's experience, there aren't any nasty side effects.) Also, when kept in glass containers in the refrigerator, micros can last for up to two weeks, and still maintain their punch.

For those who are just getting started with indoor growing, microgreens are an ideal initial project, and even for those who are experienced pros with kitchen gardens, the "crop" offers such endless variety that there's always something new and zesty to try. So, let's get growing.

## Choosing Seeds

Although most seeds can grow into microgreens, there are some choices that make more sense than others. Melons or squash, for example, produce thick and chewy cotyledons that don't taste particularly tempting. Daikon radish or purple kohlrabi, though, are both particularly flavorful and provide pretty micros. Here are some guidelines for choosing well:

*Personal taste:* Those fond of strong, spicy flavors should gravitate to mustards, arugula, radishes, cress, and other zesty greens. If a milder taste is preferred, stick with options like chard, basil, cabbage, or carrot. Imagine the full flavor of a vegetable whittled down to a sliver, and that should determine your choice. When I forgot this rule, I ended up planting a whole tray of scallion micros that overpowered every dish I sprinkled them on, and caused onion breath for three days. Unless you hate your dentist or want to ruin a first date, you may want to make a wiser choice.

*Not just for garnishes, you can prepare a whole salad of microgreens for an amazing (and amazingly nutritious) taste treat. Radish and soy bean microgreens are being prepared here.*

*Some seed companies have developed certain varieties that are best for growing microgreens, but in general, you shouldn't pay more for microgreen seeds since most seeds will grow into the micro stage.*

## MIXES VS. INDIVIDUAL VARIETIES

Some gardeners and farmers enjoy creating a visual layering system, in which micros like beets, radish, and mustard create bands of different colors. Others might mix the seeds together before planting so the micros are somewhat pre-blended before harvest. Some seed companies sell mixes, which can be significantly less expensive than buying individual seeds and blending them. Another advantage to a pre-mixed package is that the varieties will mature at the same rate, making harvest much easier and ensuring a uniform "crop."

*Blending several different types of seeds before planting allows you to harvest your mixed greens directly. Simply add some seeds that go well together, like beets, radish, and mustard, to a small jar and gently stir.*

*Germination time:* Some micros mature rapidly, even within a few days, while others might putter along and take ten days to a few weeks. Most people aren't trying to time micros according to menu planning or farmers' market selling, so germination time isn't especially important, unless you're creating a seed mix. In that case, it's usually a good idea to put fast-growing varieties with each other to prevent harvesting a tray of half-grown micros.

*Color and appearance:* Plenty of micros boast visual appeal, and planting a variety based on color can yield a tray that pops. Red-veined sorrel is gorgeous, almost cartoonish, in appearance, while red and golden beets sport brightly colored stems that seem almost like neon. For micros grown into the true leaves stage, it's fun to pick varieties with visual interest like garnet mustard, with a spray of red color against a green background, or mizuna, with its tiny, tree-like leaves.

*Price:* When choosing seed mixes or other microgreen varieties, keep in mind that anything designated as "microgreen seed" should be the same price as seeds that don't have that trendy title. If you pay more for beet microgreen seeds than you do for beet seeds from the same vendor, the price difference is probably going toward holiday bonuses for the seed company's marketing department.

# Good Varieties for Indoor Growing

Nearly any microgreen seed variety is well geared for beginning growers, and it can be tough to limit an indoor garden to just a few. To get started, here are some of my favorites (see Resources section at the back of the book for seed purveyors):

*Arugula*

*'Red Giant' Mustard*

*Arugula:* Boasting purple stems, arugula is a nice addition to any mix that's heavy on green colors, and the spicy flavor is distinctive.

*'Red Giant' mustard:* Any variety of mustard is fun to throw into a micro blend because the fla-vors will really come through; with the Red Giant, the leaves have red veins throughout, making them visually appealing.

*'Early Wonder Tall Top' Beet*

*Cress (Cressida)*

*Beets:* So pretty as they're growing, with vibrant stems depending on whether you've planted red or golden beets. They take longer to mature, sometimes 3 weeks to nearly a month, so I tend to grow them sepa-rately rather than in a mix.

*Cress:* These can be very delicate and susceptible to mold if overwatered, but when cared for properly, they grow very quickly, sometimes within just a few days. They're easy to plant intensively, and have a fresh peppery flavor.

*Komatsuna*

*Komatsuna:* A Japanese spinach with a strong mustard flavor, the microgreen version mimics the vegetable's round, green leaves.

'Ruby Red' Chard

*'Ruby Red' chard:* With a mild, beet-like flavor, this variety also stands out for its reddish-pink stems, which look awfully pretty in a salad or sprinkled over eggs. Another great chard option is 'Bright Lights,' which combines gold, pink, orange, red, white, and purple stems. Think rain-bow chard, but in a teeny tiny version.

'Red Russian' Kale

*'Red Russian' kale:* Personally, I love kale, so I tend to grow a lot of it during the year, both in micro and full-form ver-sions. This variety has a nice pop of color thanks to a pinkish out-line around the leaves.

'Dark Opal' Basil

*'Dark Opal' basil:* Herbs are always a nice addition to a micro blend, and I like a strong basil flavor. This variety also boasts a purple leaf, which is a refreshing contrast to the dominant greens seen in most micros.

# Trays, Pots, and Other Containers

In addition to nutrient density, flavor, and the ability to impress dinner guests, microgreens are remarkably easy to grow in just about any type of pot or tray. Since they don't grow to maturity, they boast shallow root systems that make them ideal for planting in an array of containers.

Sometimes, you don't even need a tray: one summer, half-dazed from a day of planting, I spilled about a quarter cup of seeds on the passenger side of my Volkswagen Beetle convertible. Since the car is awash in mud, dirt, straw, twine, and other farm-related debris, I just wiped the seeds onto the floor mat and planned to vacuum them later, during my annual fit of vehicle cleaning. Because the car gets ample ventilation and sunshine from having the top down, and the floor mats are frequently wet from mud boots and random water bottles, the conditions were ideal for microgreens, and one morning I found a beautiful indoor mat of micro radishes, mizuna, and mustards.

Given my suspicion about the toxins that might be present in automotive mats and carpeting, my little crop went to waste, sadly. But the experience led me to use more shallow trays than I'd been employing, and to increase ventilation for my indoor efforts.

An array of clever microgreen growing tactics exist, from putting a few seeds and a teaspoon of soil into bottlecaps, to utilizing old Pyrex baking dishes. But for an ideal system, I tend to prefer open-style seedling trays with drainage slots in the bottom. The slots keep the soil from collecting too much moisture, which can

*You can use practically any leftover plastic container for growing microgreens. The best ones to use, though, are foodsafe containers that are designed for holding produce.*

quickly lead to mold in a microgreen tray, even with adequate ventilation.

Also, when the microgreens reach a certain stage of growth, about three to four days before harvest, watering them from above can flatten the plants since they're somewhat delicate and have thin stems. Instead, they can be watered from the bottom (more on this later) thanks to the drainage slots.

When choosing a pot or tray, keep soil usage in mind. Because of those roots, the micros don't need the type of soil depth you'd see with plant starts or even indoor herb gardens. Save soil by choosing a smaller container, and make harvest easier with a tray or pot that's shallow rather than deeper.

An open tray can be a boon during harvest, too. When I first started planting microgreens, I tried growing different varieties in a tray with separate sections, like you'd use for seed starting. While the sections are perfect for helping vegetables or herbs establish root balls and become hearty for transfer outdoors, it was a miserable choice for

All the equipment you'll need for a large microgreens planting is assembled here.

micros when it came to harvest. Normally, I cut micros with a scissors, but because each section was so separate from the others, I had to harvest with a tiny pair of embroidery scissors. While this was fun initially—my hands looked gigantic compared to those diminutive shears and miniscule plants—the whimsy wears off fast.

## Soil Prep

Microgreens do best with very loose soil with good drainage, so I tend to use compost mixed with a little vermiculite. The most important part of soil prep is to add some water to the mix before planting, which helps to hold in moisture during the germination phase. For more on making Bossy E's Best Blend, go back and reread page 27.

When blending water and compost, go for a consistency that's like a crumbly brownie mix—then pick up a handful and squeeze. If a few drops of water come out, that's perfect. If there's a steady stream of water, it means you've made the mixture too wet, and you should add more dry mix.

It's not mandatory to premoisten the soil this way, and I've planted plenty of trays that did fine without it, but I've found that it can speed germination time by a few days if you use this method.

# Planting and Care

## First Steps

Make sure you start with a clean tray, pot, or other container, and add just a few inches of your soil mix, making sure to "fluff it up" if necessary.

Because the mix is moist, there's often a temptation to push the soil down as if making a mudpie, but this creates a soil compaction issue that can turn your micro tray into a hardened brick when it's put under light.

Here's the tricky part: seed generously. For those who are used to carefully dropping a single seed into a pot, it can take some time to overcome the psychological hurdle that microgreens present. But you need to seed heavily so that you can maximize your container's space, and also harvest more efficiently.

Granted, I'm guessing that you're not likely to become a professional microgreen grower, so harvest time is probably not a factor in your gardening strategy. But if you want to dress up a dish with a handful of micros, it will feel surprisingly time consuming to be clipping each one individually.

Here's an even more challenging part for many gardeners: don't cover the seeds with soil, vermiculite or anything else. If you do, then they tend to germinate unevenly, which isn't important if you're sowing seeds in a field, but is frustrating if your "field" is a small tray on your kitchen counter.

Prepare a blend of Bossy E's favorite soil (see page 27) and place a layer into your planting dish. If your soil seems too dense, just fluff it up a little.

Water very lightly, and then place a dish towel or empty tray over the top of the micro tray. This will help to keep the soil warm, and blocking the light for a few days helps the seeds to become healthier in general. You can peek inside if you want to see the magic—this isn't a soufflé—but be sure to replace the cover if the seeds haven't sprouted yet.

Once they show any sign of growth (about three to four days), remove the cover and water daily. Place under lights for at least six to eight hours per day.

Spread the seeds fairly densely. Don't cover the seeds with soil, vermiculite, or anything else. If you do, then they tend to germinate unevenly.

Lightly water the soil, taking care to avoid any pooling.

After a light watering, a dishtowel or other cover will help to retain heat and boost germination. Once they show any sign of growth (about three to four days), remove the cover and water daily.

## Maintaining Growth

Once they've established, microgreens don't require much care—that's one of the major benefits of growing them. But there's one important maintenance task that, if skipped, can put your micros in jeopardy.

The task is watering them properly. When they're just in the first stage of growth, before getting their true leaves, it's fine to water from above, but once they're looking very micro-like, it's crucial to "bottom water" them so the delicate stems and leaves don't get flattened.

Simply fill a kitchen sink, bathtub, even another empty tray, with about an inch of water and set the micro tray into it for a few minutes. The plant will take up the water it needs and hydrate the roots that way. This tactic is also useful if micros are looking droopy and need to perk up.

If you don't have a way to bottom water, then just water carefully at root level, or mist heavily instead of watering.

## Troubleshooting

Some common problems and potential solutions for micros:

### Moldy Clumps Inside a Tray

This can be a result of improper airflow, overwatering, or too much humidity. If there are unaffected micros in the tray, lift out that section (throw out the moldy part) and relocate into a different tray, letting the soil dry out a little before you water again. Wait a few days before harvesting to see if mold develops on this section as well; if it does, discard and start over.

*Tip: Once the greens have sprouted, do not water them from above. Instead, bottom water by setting the tray into a larger container and letting the growing medium soak up the water. Alternatively (but not as safe), water carefully at the root level or mist.*

### Tiny Threads of White Around Seeds

Fortunately, this isn't an actual issue, it will just feel that way. This isn't mold—those little threads are called root hairs, and they're crucial for plant growth. They're very common for micros, and their purpose is to collect nutrients and moisture from the soil and deliver them to the entire plant. It's easy to mistake them for mold, but don't throw out the micros thinking that you've flubbed it—they're a sign that things are going right.

### Yellow Micros that Look Sickly

If the micros are yellowish when you take off the top tray or dish towel, then you don't have a problem. They only appear that way because they need light to green up, and that should happen within a day or two at the most. If the micros are yellowing at another time, then they might be too far from their light source. Place them within 6 inches of whatever bulb you're using—unless it's a grow light that gives off heat (in which case: What are you doing using that? Are you trying to grow lizards?), don't worry about "burning" the micros, they should get a nice boost from the added photosynthesis.

### Slow or No Germination

Most of the time, this happens because seeds are too old or because the tray isn't getting enough moisture. If it's the latter, simply water more often. If it's the former, you can do a germination test by putting a few seeds on a wet paper towel. They should start to germinate within two to four days, and if they don't, then get some fresh seed and give it another try.

# Harvesting and Preservation

## Getting Ready to Pick

Grab a handful and cut. Really, it's just that easy. In some cases, you'll see seeds that haven't germinated because others have blocked their light, so you can keep the tray going and see if you can get a second harvest from it. But in general, clear the tray and then start again.

*Under good growing conditions, your micro-greens should be ready to harvest in one to two weeks. Most types are at their tastiest when they are 1 to 2 inches tall and have spouted a second set of leaves.*

*Harvesting micros is easy work, and just takes a few moments with a scissors.*

*Store freshly harvested microgreens in a glass jar with a lid. Don't pack them in tightly.*

## Storage Considerations

Although many microgreens are sold in plastic containers—at Bossy Acres, we sell them that way at farmers' markets because it's easier for transportation and cooling—the best way to keep them fresh is in glass containers. Pop them in a Mason jar or a Pyrex container with a lid, and they'll keep for a few weeks, sometimes even longer, in your refrigerator.

*In a sealed glass jar your microgreens will keep for at least a week if refrigerated. But you should always try to eat them when they are at their freshest.*

# Using Microgreens

The bad news is that long-term storage of microgreens isn't possible or realistic, unless you throw them in a pasta sauce that gets frozen or canned. But with the very short timeframe that comes with growing micros, keeping them long-term doesn't feel as important as it would for some crops.

The last bit of advice I have for micros: use them liberally. For many people (including myself), microgreens can spark creativity in terms of cooking and flavor combinations. Put them over salmon, chicken, tofu, or pork. Top a pizza with zesty arugula micros, or sprinkle some kale micros on a sandwich.

Just when scrambled eggs seem like the most boring dish on the planet, I throw a handful of Bright Lights chard micros on top and boom, I'm an amateur chef. Plus, who doesn't love a tiny forest of plants that grows within just a few weeks? Microgreens have become an indispensible part of my food landscape, and once you start growing them, I suspect they'll find a place on many of your plates, too.

*Seared Ahi tuna with fruit salsa and microgreens*

# Pea Shoots, Sunflower Shoots, and Popcorn Shoots

Sometime around the middle of February, it always seems to hit: the weariness of filling my shopping basket with fresh vegetables from California, Chile, Mexico, and even Peru or New Zealand.

It's not easy to eat local when you live in a place that requires budgeting 20 minutes every morning for scraping the ice off your windshield. The only way it can be done in even a small way is to either grow what you can indoors, or tap into whatever canning and dehydrating you accomplished the summer before.

Since we tend to run out of our food preservation goodies by March in our house, it prompted us to begin experimenting with easy-to-grow options, and that led right to shoots.

Unlike sprouts—which are grown in water and require meticulous scheduling for rinsing and care to avoid bacterial issues—shoots are grown in soil, so it's a snap to be more flexible in terms of timing, resource usage, space, and other major factors. We began growing them just to see what would happen, and now our little farm is becoming a supplier of shoots for local restaurants—one of them has even dubbed a pea-shoot-laden veggie dish the "Bossy Burger." Who wouldn't want to eat that?

With these three options, shoots are only just a few weeks away from thriving under your own lights, so let's get started.

*Shoots make a delicious garnish on seafood or meat, or you can simply eat them by the bowlful. A pea shoot salad with lemon and olive oil is a sumptuous side dish.*

# Good Options for Growing Shoots

In general, shoots are really just microgreens of the seeds you've chosen. So, if you were to let the peas, sunflowers, and popcorn grow into full maturity, you'd have those plants. But because you harvest them at such a young age, the flavors are more intense, and you can plant them much closer together in a flat tray. With shoots, variety doesn't tend to matter very much, although I've found better results with a specific type of pea seed. Here are some options:

*'Dwarf Grey Sugar' peas:* After much experimentation, we now use only these, both for our farm and for home growing. They're developed to be shorter peas, so if you want to put some in your garden, you can get pea pods without having to do trellising. For indoor growing, I like them because they're hearty, easy to grow, and taste really amazing when you turn them into pesto. Nearly any other pea seed can be turned into shoots, however, so if you have some 'Sugar Snap' pea seed at home, that will work. I'd avoid varieties designed to be tall and robust, though, such as 'Oregon Giant.'

*'Dwarf Grey Sugar' Pea shoots*

*Sunflower shoots*

*Sunflower:* Although I include a list of resources at the end of the book, I'd planned to avoid men-tion of specific seed companies in the main body of the book because I don't want to seem biased. But here I'm going to break my own rule, because there's one company that provides excellent sun-flower seeds for shoots, and they're even organic. Johnny's Selected Seeds, a company out of Maine, offers a sunflower seed variety that's highly depend-able and very hearty. When I've tried other options, it's been disappointing, especially when I've played around with growing from sunflower seeds that are usually meant for bird feeders (I mentioned gardening is an adventure, right?). Stick with seeds meant for human consumption, and give Johnny's a try.

*Popcorn shoots*

*Popcorn:* On a whim, we wondered if we could buy organic popcorn from the co-op, sprout it, and then plant it and eat it as shoots. As it turns out, the answer is yes. So, you don't need to buy "seed" from anywhere as long as you have access to pop-corn kernels that aren't processed, coated in oil, or salted in any way. Popcorn differs from corn kernels because popcorn is derived from specific strains of corn that were cultivated for the purpose. That means you might be able to sprout corn and get the same effect, but popcorn is more reliable if you're trying to grow shoots.

*Nasturtium shoots*

*Nasturtium:* One the most popular edible flowers, nasturtiums also make deli-cious shoots. As when planting in the garden, nasturtiums can be a little slow to sprout. Try soaking the seeds in water for an hour or longer before spreading them in the media.

*When planted in a shallow tray, shoots quickly develop a thickly knit root mass. It isn't edible, but it makes great compost.*

## Trays, Pots, and Other Containers

Much like other types of microgreens, I tend to prefer open-style seedling trays with minimal depth and drainage slots in the bottom. The slots keep the soil from collecting too much moisture, which can quickly lead to mold in a shoots tray, even with adequate ventilation. That's because the seeds are sown so close to each other that without drainage, root rot can become an issue within just twenty-four hours of planting.

When choosing a pot or tray, keep soil usage in mind. Because of those roots, the shoots don't need the type of soil depth you'd see with plant starts or even indoor herb gardens. Save soil by choosing a smaller container, and make harvest easier with a tray or pot that's shallow rather than deeper.

Even though peas, sunflowers, and corn require plenty of space when it comes to root depth, when planted as shoots, a very curious thing happens: the roots begin to curl up into each other, creating a thick mat in the case of pea shoots, and turning your tray into a solid block. I tend to use shallow trays just for this reason, because composting the soil after harvest is easily done, like throwing a very small carpet onto the compost pile. Bonus for homesteaders: chickens *love* the leftover stems and seeds from shoots trays.

So, you use minimal soil, get maximum use from your tray, and maybe even make your chickens happy.

# Prep Work

For shoots, I find that a pure compost mix tends to work very well, as long as it's fairly "fluffy" in consistency. If all you have is indoor potting soil, lighten it up by adding vermiculite to the soil.

The most important part of soil prep is to add some water to the mix before planting, which helps to hold in moisture during the germination phase.

When blending water and compost, go for a consistency that's like a crumbly brownie mix— then pick up a handful and squeeze. If a few drops of water come out, that's perfect. If there's a steady stream of water, it means you've made the mixture too wet, and you should add more dry mix.

It's not mandatory to premoisten the soil this way, and I've planted plenty of trays that did fine without it, but I've found that it can speed germination time by a few days if you use this method.

For seed prep, there's a simple way to speed germination time: just soak the seeds for about twenty-four hours before planting. In some cases, especially when the house feels on the colder side, I soak them for a day or two longer so that they begin to sprout before I plant them. This can reduce germination by up to a week in some cases.

But be very careful about soaking during hot weather and for too much longer than about seventy-two hours maximum, because once they pass the stage where they're sprouting, they'll start to deteriorate and can't be planted. The worst

*These sunflower seeds have just begun to sprout. Soaking the seeds in water for a day before planting them will speed up the shoots.*

part of forgetting to plant or putting it off for too long is that rotting, wet shoots seeds give off a stench that will permeate the house. Seriously, it's one of the worst stinks that you can imagine, like fungus-crusted toejam inside of wet, dirty socks. Just make it a point never to find out if I'm characterizing that aroma accurately.

# Planting and Care

## First steps

Make sure you start with a clean tray, pot, or other container, and add just a few inches of your soil mix, making sure to "fluff it up" if necessary. Because the mix is moist, there's often a temptation to push the soil down, but this creates a compaction issue that can turn your shoots tray into a hardened brick when it's put under light.

Another consideration when spreading soil is to take a moment to create a level surface, especially along the sides. Unfortunately, it's easy to let the soil build up on the sides slightly and come to a depression in the middle, but this inadvertent valley-like tray will cause unevenness in your water distribution, allowing some seeds to sit in water for too long while others dry out.

Once your soil mix is ready, seed generously. For shoots, this means you'll be creating a seed bed where the seeds look far too close together, but aren't overlapping. There's no need to poke holes in the soil mix and place seeds inside, they can sit on top of the soil and grow perfectly well that way.

Keep in mind that each seed will grow a shoot vertically, so it's fine if they're all just a few fractions of an inch from each other, but you don't want them competing for the exact same tiny stretch of soil.

If you do happen to sow them thicker and they're touching one another in parts, that's fine as well,

*Pea shoots take around two weeks to mature for eating. A) Newly planted; B) 3 days; C) One week; D) Two weeks*

*Watering shoots from beneath is a good way to hydrate the roots without flattening the tender stalks.*

just know that you might get less germination than expected. The good news is that these non-germinating seeds tend to get a second chance after your first harvest.

Water very lightly, and then place an empty black tray over the top of the micro tray. This will help to keep the soil warm, and blocking the light for a few days helps the seeds to become healthier in general. You can peek inside if you want to see the magic, but be sure to replace the cover if the seeds haven't sprouted yet.

Once they show any sign of growth (about three to four days), remove the cover and water daily. Place under lights for at least eight hours per day. You can give them a blend of artificial light and sunlight if you're in a season or a geographic area where the sun is strong enough, but keep in mind that the shoots will bend toward the light, so rotate the tray daily if necessary. Otherwise, place the tray directly under your growing lights, about 6 inches from the tops of the plants.

## Maintaining Growth

Once they've established, shoots don't require much care, which is a plus. But much like microgreens, they benefit greatly from a "bottom watering" strategy.

Simply fill a kitchen sink, bathtub, even another empty tray, with about an inch of water and set the shoots tray into it for a few minutes. The plant will take up the water it needs and hydrate the roots that way. This tactic is also useful if shoots are looking droopy and need to perk up.

If you don't have a way to bottom water, then just water carefully at root level, or mist heavily instead of watering.

## SHOOTS ARE A FOUR-SEASON CROP (BUT SOME SEASONS ARE BETTER THAN OTHERS)

When it's particularly humid in the house, (summertime in most cases) consider planting and replanting shoots only after the weather has cooled off. At Bossy Acres, we've tried to grow shoots year-round, and I've attempted the same tactic at home, but there are definitely better seasons than others when it comes to successful growing. Spring and fall are ideal. With heat mats, winter brings slower growth but still plenty of tasty shoots. Summer, however, is brutal for shoots even in a house with central air conditioning. Unless you love playing around with tweaking elements constantly—airflow, more watering, less watering, bottom watering, misting, more airflow, different placement, etc.—I'd suggest saving shoot growing for more temperate seasons rather than summer.

*Shoots grow best in the spring and fall, but if you use a heat mat or another suitable heat source you can grow them successfully in winter too. Summer, with its heat and humidity, is a less-than-perfect environment for growing shoots indoors (but that's usually okay, because summer is a great time for growing edibles outdoors).*

# Troubleshooting Shoots

Some common problems and potential solutions for shoots:

## No Germination, Even after a Few Days

Shoots tend to germinate quickly, so this would be a major issue. Most likely, it's related to the temperature of the soil. Putting a cover over the tray tends to keep the moisture and warmth inside, but sometimes you need an extra boost, so consider setting the tray on a heat mat (see page 74), which are available at any garden store. These rectangular mats raise the temperature slightly, just enough for the seeds to feel cozy, but not enough to damage them or to scorch your counter. You can leave it plugged in for days if necessary. If it's been a week and you're still not seeing germination, check the density of your soil mix and see if it's too hardened—compaction will often prevent the seeds from establishing.

## Moldy Clumps inside a Tray

Moldy clumps can be a result of improper airflow, overwatering, or too much humidity. If there are unaffected shoots in the tray, lift out that section (throw out the moldy part) and relocate into a different tray, letting the soil dry out a little before you water again. Wait a few days before harvesting to see if mold develops on this section as well; if it does, discard and start over. If the problem tends to be too much moisture from the start, which can happen during humid months, sow the seeds into dry soil instead of pre-wetting the soil mix, and do a thorough watering before covering the tray during the germination process.

## Tops of the Shoots Are Browning or Turning Dark

You have them too close to the light source, so you're essentially cooking them in the tray. Move them away from the light by another 6 inches, clip off the browned tips, and hopefully the remainder will thrive.

## Shoots Are Yellowing or Looking Droopy

Usually, yellowing means overwatering, so let your soil dry out a little. By contrast, droopy shoots are often caused by underwatering, or by too much humidity. In that case, bottom water the shoots in cold water and see if they perk up after about ten to fifteen minutes. Also, adjust your fans so that shoots are getting some airflow to disperse stagnant, humid air and increase circulation. Keep in mind that plants "breathe," and they need some airflow to keep oxygen levels high. If your efforts aren't successful and the shoots remain droopy, you may just want to harvest and eat—they'll still be tasty and nutritious, just a bit less crisp.

## TIPS FOR GROWING SHOOTS

Here are some other tactics for keeping your shoots on track:

Continue watering and providing light to the shoots tray after a first harvest. Many times, some seeds germinate later or are delayed because they're under other seeds. For pea shoots especially, you can get two to three more harvests because of this. These subsequent harvests won't be as abundant as the first ones, but they're enough to be worth the effort.

If the air in your house seems especially dry, mist the plants when they're in an early stage of growth and then put a clear tray on top. These are available at any garden store, and they help to let in light but lock in moisture. I find these invaluable in the winter, especially when the shoots are just out of the germination stage, but not established enough yet. Alternately, you can also use plastic wrap, but just be sure that the shoots aren't grown enough to bump against the top of the wrap, which can impede growth.

Another good tactic when the air is dry is to mist the plants more often during the day if they seem to be struggling. Be aware that this strategy could cause browning if the plants are close to an artificial light source, so raise the light or lower the plants by 6 inches and mist heavily if this happens.

*Provide a little drainage by placing small rocks in the bottom of your planting vessel prior to adding the dirt.*

Plant shoot seeds densely. Often, a second crop will pop up from the original seed planting after you harvest the first batch.

# Harvesting and Preservation

## Getting Ready to Pick

For each variety of shoots, there is an ideal time for harvest:

- Pea shoots: These will grow long and look unruly, and I tend to harvest them when they're about 8 to 10 inches. If you want sweeter and more uniform shoots, harvest at around 4 inches when they're just beginning to unfurl their leaves. If you want a super abundant harvest, and a less pea-sweet taste, let them go until 12 inches. Longer than that, though, and they start to get too long and heavy for the tray, so they start flattening. Also, the taller they get, the more woody the flavor.

*Pea shoots make a tasty and visually appealing garnish for most seafoods, including these seared scallops.*

*You can harvest pea shoots at any stage, but they tend to be best at 8 to 10 inches, before they get unruly. Younger pea shoots, like the ones seen here are just a few inches tall, but they can taste sweeter and have a more uniform size.*

## STORAGE CONSIDERATIONS

Just like microgreens, shoots do best stored in glass, and can last for weeks in the fridge this way. Pea shoots, though delicate, are remarkably long lasting and keep their crunch up until the end. Sunflower shoots are even tougher, and I've kept some for three weeks in the fridge without any discernable change in flavor. Popcorn shoots, though, tend to last for a shorter time and when kept for longer, that odd aftertaste gets stronger. So for those, I would harvest and eat the same day.

- Sunflower shoots: These taste best when you get them before the sunflowers develop their first true leaves. At the shoot stage, you'll have just two petals that are nicely crunchy. When the first true leaves begin sprouting, they come up between these petals and have a scratchy texture that I don't enjoy even though they're perfectly edible. If you let the shoots go for too long and the leaves start coming up, you can still enjoy the shoots-only taste and look by plucking off the true leaves and discarding them. In terms of what to harvest, I clip all the way down to the soil, because the stems are just as tasty as the leaves.

- Popcorn shoots: These are best at about 6 inches, and be sure to include some of the white-and-pink bottoms, where the most intense flavor resides. Once the shoots get taller, they're fun to have in the windowsill because they look so vibrant, but the taste becomes more and more grassy as they grow. At about 10 inches, the shoots have all the flavor and appeal of fresh lawn clippings.

# Herbs

I've seen some great combinations put together—like creating a "pasta sauce herb garden" with oregano, thyme, and basil—and because most herbs thrive from frequent harvest, they're perfect for a kitchen space since you can just snip what you need for a particular dish.

For my kitchen space, I like to put herbs in separate pots so that I can isolate them in case of insect issues or mold problems. But I often create combinations when giving herb pots as gifts, and I always love the aromas that blend as a result of letting them tangle up together.

In general, many herbs are well-geared toward indoor growing, and several of the gardeners I know repot their outdoor garden herbs as the weather here gets crisp, so that they can extend the growing season indoors.

Some herbs, such as rosemary, oregano, thyme, and sage, propagate well if you take a cutting from an existing outdoor or indoor plant and prepare it for growth inside. If that's your strategy, simply cut a 4-inch section (measured from the tip of the stem/leaf toward the soil) and strip off about an inch or so of the lower leaves. Put the stem into a potting mix, such as vermiculite, and keep the mix somewhat moist as the plant establishes.

These plants like humidity, so cover with clear plastic or glass—letting light in, but trapping moisture—but don't let them get too hot from direct sunlight. Also, remove their covers occasionally or put them on a porch or "transitional" space to give them some air.

This technique also works well for transplants purchased at a greenhouse, which should already be "hardened" to temperature variations. In my experience, there are times when I just can't seem to grow certain herbs from seed (I'm looking at you, basil) and I need the jumpstart that a transplant can provide. When I began my indoor growing adventures, I relied on transplants almost entirely because I appreciated being able to skip that first big leap in germination. These days, however, I actually enjoy the planting process, and tend to choose herbs that thrive best in my kitchen.

So, for this section, we'll take a closer look at how to go from seed to harvest, so you can get a feel for growing herbs entirely indoors. If you do go the transplant or herb-cutting route, though, much of the information still applies, particularly when it comes to issues like soil drainage and temperature.

*Some herbs can be started from seed indoors; others can be propagated from cuttings you take from your outdoor plants.*

# GOOD VARIETIES FOR INDOOR GROWING

Although many herbs do beautifully in an indoor growth space, some varieties can be surprisingly fickle if you're trying to grow them from seed. Here are some options, broken down by level of difficulty:

## EASY LIKE A SUNDAY MORNING

*Peppermint and spearmint:* Hearty and wide-ranging, mint likes to invade the territory of other plants, so once you've got it, stay on top of harvesting it. If you want a strong mint taste without growing a bunch of it, opt for peppermint, since it has a more intense flavor.

*Vietnamese coriander:* This variety is considered easier to grow than other varieties of coriander, and it has just as much flavor. Also, it's quite hearty and can last for months.

## TAKES EFFORT, BUT GENERALLY DOABLE

*Lemongrass:* You don't even plant this one from seed; just buy a stalk of lemongrass from a grocery store or farmers' market, trim the top, and put the stalk in a few inches of water. The stalk will produce roots on its own and dozens of new shoots, and you can harvest from these.

*Parsley:* Usually fairly easy to grow, but germination can be hit and miss. Usually, you'll begin to see growth about two weeks after planting, and it tends to grow slower than other herbs in general.

*Oregano:* The trick with oregano is giving the plant enough light every day; it may require placing the pot under a separate bulb that's on for a few hours more than the other herbs. Usually, about eight hours of light is best.

*Rosemary:* I've found that my best indoor rosemary comes from cuttings off rosemary in my garden, but it's also somewhat easy to grow by seed. Watch out for overwatering, since rosemary tends to prefer drier soil, and be sure to choose a variety that does well with indoor growing, like 'Blue Spire.'

*Thyme:* This herb also requires more light, so I often put oregano and thyme in the same space or place them next to each other in a south-facing window.

*Chervil:* Fairly uncommon, but quite delicious, chervil is related to parsley and has a sub-tle flavor. The herb does well in low-light areas, making it a nice choice for kitchen corners and out-of-the-way spots, but keep in mind that it doesn't do well when temperatures begin to rise beyond 70 degrees, so gauge your house's temp before planting.

## MORE CHALLENGING

*Basil:* It's so ubiquitous in cooking, you'd think basil would be a snap to grow in your kitchen garden. But no. Notoriously difficult to grow indoors from seed, basil tends to work best as a plant start from a greenhouse. Also, those lush Italian basil leaves may not be as wide and pretty as you find in a garden. Instead, I lean toward varieties with smaller leaves like 'Dark Opal' or Thai basil.

*Sage:* It isn't too difficult to get sage started indoors, but it's prone to death by overwatering. There have been many times where I see those tough leaves and think the plant looks dry, and then I end up killing it because it's already well watered. Also, many experts advise waiting a long time before harvesting while the plant gets established—up to a year in some cases. For me, the space in my kitchen growing area is too limited to nurse a plant that takes months before I can use it, but I do like to have a small pot going because the herb dries so well after harvest.

*Cilantro:* Here's another one that grows remarkably well outside but requires a higher level of care indoors. First of all, it doesn't transplant well, so it needs to be grown from seeds or starter plants. Also, it requires plenty of drainage and yet needs more nutrients, making it tough to keep the soil nourished. It tends to do fine once it's established, but until then, plan on giving the herb fertilizer biweekly, which will be twice as much as your other herbs. Also, water only when the soil seems very dry.

# Soil Prep

Herbs are quite finicky when it comes to drainage. Common gardener wisdom is that they "don't like wet feet," which means that if their roots get too soggy, rot will result. Every plant type covered in this book (except sprouts) relies on proper drainage to some degree, but for herbs, it's particularly crucial because they tend to like a more humid environment, making them susceptible to root rot.

Simply using a pot with holes in the bottom isn't enough, and if you buy a standard type of pot with a dish that catches water, you'll be in even more trouble. This often leads to an herb sitting in water, which is almost always a terrible situation for the plant.

There are a number of strategies that can be helpful for increasing drainage. Some gardeners use a system designed for growing cactus, because these specialty mixes are designed to drain quickly, but I find that simply mixing sand and vermiculite (in a one-to-four parts ratio) together tends to make a happy blend.

If you're transplanting from outside, do as much as you can to remove the existing garden soil by gently shaking or tapping the roots. You won't be able to eliminate all of the soil, but if you can get most of it, you'll significantly reduce your expose to garden pests and diseases.

Whatever you use, a good strategy to prevent compaction is to cultivate within the pot every month or so. You can simply take a fork and gently loosen the soil within the container, taking care to stay mostly on the periphery so you don't damage the roots.

*If you are planting herbs from seed, make sure the soil is not too densely packed. If need be, cultivate it by loosening the soil with your fingers.*

# Trays, Pots, and Other Containers

At the risk of harping too much on drainage, I'm going to emphasize the importance of drainage again. I can even put it in italics if it helps: *drainage.* Herbs just don't do well at all in any situation where they're sitting in water, so choose a container or pot that allows for ample flow.

Sure, you can try growing cilantro in your middle school lunchbox or parsley in that repurposed desk drawer or whatever other funky container appeals to you, but if you want to keep the herbs going for a nice long time, drill some holes and maybe even throw some pebbles in the bottom. Your herbs will thank you.

Beyond that, any material works, but I tend to shy away from terra cotta pots because they make the herbs dry out faster, and that throws off my estimation of how much watering they need.

Because of limited space, I gravitate toward smaller pots, but if the herbs begin to spread too far over the edges, or I want to encourage them to get bigger, I'll repot into a larger container.

# Planting and Care

## First Steps

Get your container and soil mix ready (see pages 19 and 27), and give the mix some water so it's slightly damp. This will keep your seeds from shifting when you first water them.

Sow the seeds about one to three times deeper than the size of the seed. This is a general rule, and what it usually means for me is that if it's a teensy little seed, I'll barely press it into the soil, and then I'll cover it with just a sprinkling of vermiculite. If the seeds are larger, I might press about half an inch and then cover it.

Water lightly, and then cover the pot or container with plastic kitchen wrap. This will keep the soil mix and seed warm, to encourage germination. Put the pot or container in a sunny area or under a light, and when the seedlings emerge, remove the plastic wrap.

*Herb seeds don't need to be planted too deep, just press them gently into the soil with your finger.*

Fish fertilizer is available from any garden store, and when diluted, can provide a nice boost for any type of plant.

## Maintaining Growth

Some herbs require specialized care, but in general, most can benefit from these tips for keeping herbs going strong:

- Fertilize every 10 days or so with diluted fish fertilizer, found at any garden store. In a pinch, I've also soaked seaweed in hot water for a few hours, let it cool, and then sprayed that on the plants. But for this to be an "in a pinch" tactic, you'd have to be one of those people who happens to have dried seaweed in the cupboard.

- Herbs like humidity, but it's not always easy to tweak this condition, especially in the winter. One good tactic is to create a tray of small rocks or pebbles and fill the tray with water, leaving about ¼ inch of the top dry. Put the pots on top of the trays—making sure they don't touch the water—and the evaporation process will help to keep the air at a nice humidity level.

- Give herbs a regular "bath" by misting every few days. Not only does this help to keep them hydrated, but it also cuts down on insect issues since weakened plants are more susceptible to pests like aphids and spider mites.

- Water from the base of the herb, not the leaves. This will help the plant get hydrated, without subjecting it to "flattening out" from too much water hitting the leaves.

# Troubleshooting Herbs

Some common problems and potential solutions for herbs:

## Slow Growth or Limited Germination

Potentially, this could be a light issue. Herbs need at least five hours of light per day to stay healthy, and full-on sunshine (provided the area isn't too hot) is ideal. But in the winter, even a south-facing window might not be enough. If you're seeing sluggish growth, try extending the time that the herbs are under light, up to fourteen hours if necessary.

*Some herbs, like sage, are prone to browning or yellowing of the leaves. Often, it is caused by overexposure to heat.*

## Brown Patches or Withered-Looking Leaves

If your herbs are under a light source, they may be too close to the bulb. Basically, you're burning them. Herbs should be at least 6 to 8 inches away from the light source, and that distance is measured from bulb to the top of the plant.

## Whitish Fuzz on Soil

This might be natural seed germination, but it could also be mold. Check that you're not overwatering—if you put your finger into the soil mix and feel moisture immediately, then this could be the issue. Let the plant "dry out" and isolate it from the other plants in the meantime. Remove the moldy areas, but be aware that you may need to start over with this one.

## Stems Seem Soft or Mushy

This may be another incidence of overwatering. Stems tend to feel too soft when root rot is occurring. Another big indication of problems is a foul smell mixed in with the pleasant herb scents.

## Bug Infestation in Progress

It's so disheartening to lean over a pot of herbs, ready to snip off a few selections for dinner, and see that something else is ahead of you in the buffet line. If this happens, isolate the plant and try a soap spray—this can be a mix of mild liquid soap (like Dr. Bronner's) and water. Spray in the evening, since application when the plant is in full sunlight can cause drying.

# Harvesting and Preservation

## Getting Ready to Pick

Each herb is harvested in its own way, but in general, go for the "older" leaves that are full. Look for new growth, usually near the center of the plant, and avoid clipping near that area since you don't want to shorten the lifespan of the herb.

If a plant begins to flower, snap off the flowering part as soon as you can to prolong the herb's timeframe—like many plants, herbs begin flowering as a signal that they're done growing (it's a process called "bolting"). By removing the flowers, you can essentially trick the plant into sticking around for longer.

When harvesting, keep in mind that stems tend to have abundant flavor as well. For example, cilantro stems are just as intensely flavored as the leaves.

*INCORRECT: Do not harvest the new, inner growth of the herb plant.*

*CORRECT: Cut older, mature leaves on the outer area of the plant.*

*Herbs can be bundled and dried hanging upside-down right in your kitchen.*

## Storage Considerations

There are so many ways to make sure that your herbs can last for months. The quickest method is through drying—simple create a bundle, strip off about an inch of leaves to expose only stems, tie with twine or a rubber band, and hang up in a cool, dry area. There's a section of my basement that smells pretty amazing at this point, thanks to multiple herb bundles. The advantage of drying is that you can leave the bundles hanging for quite a while, or just crumble them when they're completely dry and put them into glass jars.

Another storage option is preservation in olive oil. Make sure the glass jar is very dry, and then fill it with herbs like basil, thyme, and oregano, as well as other ingredients like garlic. Fill the jar with olive oil, making sure the oil completely covers the herbs, and in about six weeks, the oil will be infused with flavor. After that, you can strain out the herbs and drizzle the new creation over whatever you like.

Similarly, you can finely chop herbs, put them in a glass jar, and cover completely with honey. In six weeks, the flavor will infuse the honey, and in this case, I don't try to strain out the herbs. This blend can be used in place of jam on toast, or some people simply eat a spoonful when they're feeling energy depleted, especially if more medicinal herbs are used.

*You have to make a significant commitment of indoor real estate to grow enough basil to keep your kitchen stocked in pesto. But if you love fresh pesto you'll find the space.*

# PART THREE

# Countertop Crops

Many plants with a shallow root system can grow well indoors, such as beets, radishes, some varieties of carrots, lettuces, and with the right conditions, even hot peppers and tomatoes.

With these types of crops, it's even more important to establish good growing practices as outlined in Part One of this book—proper airflow, a robust indoor soil mix, good lighting, pest prevention, and appropriate containers.

With a designated growing area set up correctly, the fun can begin. Although growing mini-crops like pea shoots is a jaunty endeavor, I find deep satisfaction in nurturing indoor vegetables that take time to develop, rewarding me with a plate filled with lush salad greens, sweet carrots, peppery radishes, and other bounty from my kitchen garden.

Especially in the spring, when I'm craving fresh greens but only find West Coast (or internationally shipped) options in my grocery store, I feel inspired to go on a planting frenzy, filling up every available space with shelves, lights, and just-seeded pots.

Not every vegetable is perfect for my kitchen counter, though. I've tried many varieties that might start out fine but lack the soil depth needed, or that thrive better outdoors in the field.

Here, we'll cover the different varieties of crops that make the most sense for indoor growing, and take a look at special considerations in terms of soil, pest prevention, containers, and lighting. Much like the preparation for faster-growing plants like microgreens and pea shoots, successful growing and harvesting of indoor crops is all about good upfront work, paying attention, and maintaining that all-important sense of adventure.

# Lettuces

Even before I started farming, I was a salad kind of girl. That label evolved over time, though: growing up, "salad" was a wedge of iceberg lettuce with ranch dressing and little else (we'll sidestep the fact that in Minnesota, "salad" can also be Jell-O and Cool Whip), but eventually I made my way to a huge variety of greens, including red oakleaf, butterhead, bibb, and romaine.

I love the blend of textures and colors you can put together with salad mixes, especially with choices like the 'Freckles' variety that grows with deep splashes of crimson or 'Rhazes,' which forms tight heads of red leaves that hide a solid lime green center.

*A lettuce bowl isn't just a way of serving greens at the table: it is also a way of growing replenishable lettuces indoors.*

# Get Ready

## Good Varieties for Indoor Growing

At the risk of oversimplifying the range of lettuce choices, I'll make a general claim: lettuce tends to come in one of two versions—either you have head lettuce that's true to that definition, or you have salad mixes that combine a number of varieties and grow in a cluster rather than a tightly contained head.

That doesn't mean that they're two distinct types—head lettuces can be included in a mix, you just harvest them when they're still petite greens. The difference comes during the seeding process, because you either space them to accommodate for head lettuce growth in those varieties, or you seed liberally (though not as heavily as microgreens) so that you can harvest several varieties simultaneously.

When it comes to indoor growing, I nearly always gravitate to the mixes because they grow quickly, I can seed them according to pot size, and they incorporate a range of flavors from mustard greens and peppery arugula to pretty 'Rouge d'Hiver' and frilly 'Lollo Rosso.'

Many seed companies sell their own mixes, but you can also create your own from contenders like these:

*Arugula*

*Arugula:* Also known as rocket, this is a dependable choice for adding a peppery flavor to salads. The spiky, dark green leaves mix well in a salad but also can be substituted for spinach in some dishes.

'Red Oakleaf' lettuce

*'Green Oakleaf'* and *'Red Oakleaf':* Not surprisingly, these lettuces get their name for having leaves that are similar to those of oak trees. In this case, *oakleaf* is an umbrella term with several varieties in that family, such as 'Tango,' 'Bolsachica,' 'Panisse,' and my favorite, 'Sulu' (which I love because I'm a sci-fi nerd). Oakleaf varieties are very easy to grow, and tend to have a mild flavor and crunch.

*Baby leaf lettuce (Romaine)*

*Baby leaf:* Much like oakleaf, "baby leaf" is a general term rather than a variety in itself. Designed to be harvested at an early stage of growth, some nice picks for baby leaf lettuce are 'Red Sails,' 'Refugio,' 'Parris Island,' and 'Defender.' Here's a head's up for seed ordering: some baby leaf options are also oakleaf varieties, but not all oakleaf choices are designated as baby leaf. Confused yet? Don't worry—the designation doesn't matter for growing, it's just a way to indicate which lettuces are best when harvested at a petite stage.

*'Lollo Rosso'*

*'Lollo Rosso':* Incredibly frilly and slightly bitter, this lettuce has an intense flavor that gets stronger as it gets larger. The red leaves have a bright green splash near the bottom, and it tends to be best when mixed with milder lettuces rather than served solo.

There are so very many varieties of lettuce that it's really a matter of taste and visual preference as to what you pick. I tend to like a bit of peppery flavor, a hint of bitterness, and an array of reds and greens. If you like only mild greens, opt for a single variety that promises that color and taste. If you can't decide and just want numerous seed varieties blended together, then go for one of the lettuce mixes offered by the seed companies that specialize in those.

No matter what you pick, I'd advise jotting down growing notes in a journal. When growing several varieties of lettuce, I sometimes forget what works well and what doesn't, or what tasted too bitter or too boring. A quick perusal of my gardening journal before ordering always helps me to distinguish the bountiful from the blah.

## Trays, Pots, and Other Containers

Because lettuce has a shallow root system, it works well in a medium-sized container. I tend to use hanging baskets because I can move them around more easily, and because they look nice as a focal point in the kitchen, especially if I'm growing lettuces with plenty of red leaves.

Like other plants, lettuce will do better in plastic than in terra cotta pots, because the clay will dry out the soil mix faster than the plastic. If you like the look of terra cotta, just pop a plastic insert into pot. Just make sure there are holes in the bottom for proper drainage.

*Plastic planters dry out less quickly than clay and tend to work out better for just about any plant, indoors or out. The oblong shape is perfect for a short row of lettuces.*

## Soil Mix

As long as you're not using soil from your garden, lettuces can thrive in a variety of soil types. You don't have to play around with drainage options by including rocks, or mixing sand into a compost blend—just grab a bag of all-purpose potting mix and you're good to go. Ideally, the mix should be organic and designated for vegetable growing.

# Planting and Care

## First Steps

Fill the pot with soil, with about an inch of space between the top of the container and the soil. Sprinkle the lettuce seeds on top of the soil, taking care to separate any that are too close to each other. For a medium-sized pot, you'll probably use about twenty seeds, but it's not necessary to be picky about it and count them out. You don't need to get anxious about spacing, either, as long as they aren't clumped together.

*Sprinkle potting soil lightly over the top of the seeds, only enough to barely cover them. If the top cover of soil is too heavy, it will prevent light from boosting the germination.*

*Spray the seeds with water from a mister bottle so that soil is moist but not thoroughly soaked. Watering them directly from a watering can could cause the seeds to shift or to be driven farther down into the soil.*

## Maintaining Growth

Once the lettuces begin to establish, there are several tactics for fostering growth:

- Water every other day to keep the soil moist but not soaked. The way to tell if it needs water is to put your finger about half an inch into the soil and if it feels dry, then water.

- As lettuces get larger, it's usually better to bottom water. This means filling a sink or tray with a few inches of water and placing the lettuce pot into it for about ten minutes. Don't let the lettuce sit in water for a long time, or collect water in a plate under the pot; this can lead to root rot.

- You may notice that several lettuces are sprouting from the same area. Weed out the weaker seedlings so that the stronger ones can have more room to expand. If the weeded-out seedlings look viable at all, just transfer them to a separate pot or eat them as microgreens.

- Fertilize if growth seems slow. I don't usually use fertilizer for lettuces because they don't tend to need the boost, but if your lettuces are looking like they could use some extra nutrition, apply a fertilizer once a week for three weeks and see if that makes a difference. Don't apply more than that, though, or you could "burn" the plant through over-fertilization.

## Troubleshooting

### It's Been a Week and There's No Germination

If you've been watering faithfully and giving the lettuces enough sunlight, then be patient and wait another week; sometimes, in certain conditions, they can be sluggish in terms of germination but will establish fine once they start growing. If it's been a few weeks and there's no germination, you may be overwatering or setting the pot in a location that's too dark.

### The Lettuces Seem Droopy

One of the nice things about living in the Midwest is that lettuces do well indoors because so much of the year is cool (maybe a little too much of the year for some folks). Lettuce loves cooler temperatures and will thrive most at around 60 to 70 degrees Fahrenheit. If you're experiencing higher humidity or temperatures in your house, try moving the lettuce to a cooler location.

# Harvesting and Preservation

## Getting Ready to Pick

Harvesting lettuces is a breeze—just snip them at whatever stage of growth appeals most to you. Usually, I prefer smaller lettuce greens rather than larger leaves, so I cut them when they're only about 6 inches high. No matter what size you choose, just be sure to avoid the inner leaves of each lettuce cluster, since it contains the immature growth that will lead to your next salad.

## Storage Considerations

Like microgreens, lettuce tends to keep best in glass containers. They also do fine in plastic bags placed in the crisper drawer of the refrigerator. In general, though, it's better to harvest only what you need for a particular meal and let the plant keep generating lettuce leaves.

*(RIGHT) Whether it's a single plant or a whole row, lettuce grown indoors brightens the house and is a very convenient source of greens.*

*(BELOW) Choose lettuce leaves on the outer part of the plant for picking. Let the inner ones stay so the plant continues to grow.*

# Radishes

Although I've always included lettuce in my food mix, radishes are a recent addition. In fact, despite significantly boosting my culinary expertise over the past decade (it's not hard to get better at cooking if you start out eating potato chips for dinner—from that base, anything is an improvement), I only began eating radishes when Karla and I grew them at Bossy Acres.

Radishes were our first crop to harvest, and I was shocked at how flavorful they were. I'd always believed that radishes were dry and peppery, based on the reactions of other people who said they hated the vegetable. But freshly harvested radishes—particularly those that aren't left in the field for too long—are actually somewhat juicy, and impart subtle flavors. Some have a kick, true, but those can be tamed with quick pickling (see storage notes later in this section).

For indoor gardeners, radishes can be a fun crop to grow, because they mature very quickly, sometimes in only about three weeks from seeding. You can eat the greens, and their bright colors (for most varieties).

*The best thing about growing radishes (indoors or out) is that they grow very quickly; sometimes they're ready to eat less than a month after planting.*

# Get Ready

## Good Varieties for Indoor Growing

Because radishes are a root crop, you'll want to pick varieties that will fit in whatever container you choose. If you want to experiment with growing a long radish in a tall container, you could opt for a Daikon, which can grow up to 18 inches long and usually takes two months to mature. In general, though, I tend to like fast-growing varieties that don't take up much space. Here are some options for those:

## Trays, Pots, and Other Containers

When choosing a container, you can plant several radishes in a round pot, but I tend to like a narrow, rectangular pot to mimic the way that radishes would grow in the field. They need to be at least a few inches apart for adequate growth, so putting them in a long container will create a nice row that's also visually appealing.

## Prep Work

Consider planting radishes during a cooler part of the year, like late spring or early autumn. Although radishes planted indoors won't be as sensitive to sudden temperature changes the way they would outside, they do like cooler temps. That's why they're often one of the first crops to be harvested at the farm.

In terms of soil, standard indoor potting mix will work, and just make sure that there's adequate drainage. This is one case where a little compost mixed into the soil would be beneficial, since radishes tend to like some added nutrients.

*French Breakfast radishes*

*'French Breakfast'*: More tapered rather than round, this is a popular variety for its white-tipped end and crunchy texture.

*'Cherry Belle' radishes*

*'Cherry Belle'*: These mature in less than a month and tend to have a mild flavor. True to their name, they have a bright, cheery red exterior.

## Planting and Care

### First Steps

Sow the radish seeds about ½ inch deep, which usually means putting them in the soil and poking them down slightly. I like to place the seeds about 2 inches apart, with the recognition that I might need to "thin" them at some point to let the heartier radishes thrive.

Sprinkle some potting mix on top of the new seeds, and water so that the soil is dampened but not soaked. Place in a sunny spot, or under a full-spectrum fluorescent. To help with germination, create a "mini-greenhouse" environment by covering with plastic wrap for a few days until you begin to see them sprout. Once this starts, remove the wrap and mist with a spray bottle until the soil is moist.

Sow the radish seeds so they are ½ inch deep and about 2 inches apart.

After planting and topdressing with a light layer of potting mix, water the radishes and then stretch some plastic wrap over the pot for a couple of days.

## Maintaining Growth

Here are some tips on getting your radishes from seed to harvest:

- Water regularly—unlike some crops that should be watered when the soil is particularly dry, radishes do well with a frequent, predictable watering schedule. This will help them to grow more quickly, which leads to a milder taste rather than the overly peppery, woody taste that occurs when radishes grow too slowly or are left for too long before harvest.

- Plant more in additional containers every few weeks—this is equivalent to crop rotation, but on a tiny scale. Planting on a continual basis will ensure that you always have radishes in progress to replace the ones you harvest, so you don't have to wait another month after picking to get more radishes.

## Troubleshooting

Some common problems with radishes and potential solutions:

### Leaves Are Brown and/or Drooping

Most likely, this is a light issue. Since radishes prefer cooler environments, try moving the pot to an area with more indirect sunlight, or moving it farther from its grow light.

### Radishes Don't Seem to Have Enough Room to Grow

In the field, we plant radishes quite close together and then thin them later, so that the strongest are given the chance to thrive. This increases the yield on a harvest, but for a home garden effort, you don't need to plant them quite this intensively. If you do happen to experience crowding, simply pluck out the young radishes that seem weaker, and remember, you can eat the greens so that they don't go to waste.

### My Radishes Are Tough and Taste Dry and Woody

There's not much that can be done at this point since they've already been harvested, but it's good to note for the future so that you can harvest them earlier next time. Some people believe radishes should be larger before they can be eaten, so they wait until the radishes are the size of small apples, but for most varieties, that's too large.

*Pick radishes when they are still relatively small and tender.*

# Harvesting and Preservation

## Getting Ready to Pick

Often, much like carrots, radishes will "shoulder" when they're ready to be picked. That means they push up out of the soil with part of the vegetable showing. This is hugely helpful in determining the size of the radish, which should be about an inch in diameter at harvest time.

If they aren't shouldering, gently dig into the soil with your finger until you can feel the top of the radish, and this will give you an idea of the size. If they're not ready, just replace whatever soil you've moved and check again in a few days.

Once they're ready for harvest, just take hold at the base of the leaves and give a firm tug. It doesn't take much effort, and they tend to pop out easily.

*Radishes are ready for picking when the shoulder is about an inch across—don't let them get any bigger, as they tend to get woody.*

## Storage Considerations

If you'll be storing radishes instead of eating them immediately, just brush the soil off and store them instead of washing them right away. This helps to keep them fresh for longer.

Radishes last for weeks if stored in a plastic bag in the crisper drawer, but my favorite tactic is to create a quick pickle that doesn't require any canning equipment; you simply store it in the refrigerator. This is the only recipe I'll include in the book, because during the farm season, it's what I make on a weekly basis, and I feel like it really brings out the flavor of the radishes.

# Carrots

To be honest, carrots grown in containers can be tricky, but you can boost your chance of success by choosing the right varieties. My favorites in the field tend to be rainbow carrots, with their long tapers of pinkish red or dark crimson, both cut open to reveal a rich orange inside.

As much as I'd love to plant those type of full, luscious beauties inside, however, it just doesn't make much sense to utilize that much soil and lug around a large container just for the thrill of growing larger carrots indoors. I get my fix instead from smaller varieties, and also from growing just one or two pots at a time.

Would I like to turn my office space into a carrot-filled zone, with radishes and hot peppers on the periphery? Of course I would. Maybe someday I'll decide never to sell my house and start building raised beds in less-used spaces. I'll be that weird lady who gets raided by the DEA because a neighbor sees massive grow lights in every room.

Until that time, though, let's go with carrots that make sense for a small-scale growing plan.

*With their intense flavor (compared to grocery store carrots), carrots grown in-home go a long way when seasoning soups.*

# Get Ready

## Good Varieties for Indoor Growing

Resist the temptation to grow your own juicing garden (unless you actually do have an indoor greenhouse, and in that case, full speed ahead to you) and opt for varieties that are more round than tapered. These work very well:

'Nantes' family carrots

'Parisienne' carrots

*'Parisienne':* Looks like radishes, tastes like carrots. This variety is easy to grow and if there's such a thing as an adorable carrot, it's this one.

'Tonda di Parigi': Another choice for a radish-size carrot that resembles the 'Parisienne' carrots above, this one takes just sixty days and only grows about 2 inches long.

Another option is a type of carrot called 'Nantes' that has the tapered shape that's familiar to all carrot lovers, but is shorter and wider than traditional varieties, making it better for indoor growing. Try these:

*'Shin Kuroda':* A Japanese variety that takes about seventy-five days to mature, but results in a tender carrot that's only about 3 to 5 inches long.

*'Danvers Half Long':* If you have the container space for a 7-inch carrot, this is a nice choice since it's dependable and hearty.'

*Little Finger':* Although these take a bit longer to grow, they only grow about 3 inches long, which make them perfect for indoor gardens.

## Trays, Pots, and Other Containers

Even with shorter varieties, carrots need ample room to grow, so use containers that are at least a foot deep, if not more. Even though the vegetables won't take up that space, their root system will use the room to develop properly.

You can opt for a rectangular container to mimic the look of carrots in a field, or utilize a round pot, which looks cool when all the carrot greens are fluffed up together. As with other crops, terra cotta pots are less ideal because they tend to draw moisture from the soil, but with a regular watering schedule, this is less important than it would be for something like lettuces.

When selecting a planter, make sure that it's thoroughly cleaned if you've grown plants in it previously. Even with old pots, it's possible to transfer diseases onto the new crop, and since carrots are root vegetables, they would be more affected by any contaminants that leach into the soil.

## Prep Work

Carrots do well in loose soil, and even a small degree of compaction can result in stunted growth. Because of this, you might consider a soilless mix like compost and sand mixed together, or at least putting some vermiculite in your potting mix to allow for more drainage.

This may be a chance to try out coir bricks, if you haven't played around with those before. These bricks, made up of coconut fiber (called "coir") can be expanded to create a carbon-rich base beneath potting soil, or parts can be broken off and used to create a looser soil mix. As an added bonus, coir is great for compost piles, and is often used for vermiculture, so if your next step in gardening is creating your own worm bins, coir should be on your greenhouse shopping list.

*Coir bricks are made of compressed coconut fiber. You can break off pieces and mix them with compost for a perfect carrot-growing medium.*

# Planting and Care

## First Steps

Like radishes, carrots tend to do well when planted during cooler times of year like late spring or early fall, but in an indoor environment, they can thrive as long as the temp doesn't get too hot.

After creating your soil or soilless mix, fill your container and leave an inch of empty space at the top. Poke your finger into the planting medium where you'd like the carrots to go, about half an inch deep. Space the seed holes about 3 inches apart, if not more.

Drop two to three seeds into each hole, which will increase your chances of having at least one sturdy, hardy carrot in each spot. If more develop, you can simply thin them out and let the strongest one keep growing. Sprinkle soil or vermiculite on top of the holes, being careful not to press down when filling, since compaction could hinder germination.

*Poke ½-inch deep holes for the carrot seeds, about 3 inches apart.*

Water thoroughly, so that the planting medium feels very wet but not flooded. Place pot in an area that gets at least six hours of light per day, usually in an area with some direct sunlight and some indirect. Carrots do well in shade, but also grow better with at least some partial sunlight.

*Sprinkle some soil or vermiculite on top of the planted seeds, then water. Continue to water regularly until the carrots are ready for picking.*

## Maintaining Growth

Some tips on getting carrots from seed to harvest:

- Keep the soil or soilless mix well watered. Because you've created a looser blend than for other vegetables, the carrots will probably need to be watered more often, especially during hotter days. Carrots can tolerate some dryness in the soil, but prefer a steady supply of moisture for better growing.

- If growth seems sluggish, you can apply fertilizer once a week for a few weeks, but this isn't necessary if growth is proceeding well.

- Thin out the carrots if they begin to seem crowded by simply plucking out the greens that are smaller than others. Be careful during the process, though, since you don't want to damage the roots of the remaining carrots.

## Troubleshooting

Some common problems with carrots and potential solutions:

### Greens Are Flopped Over into the Pot

Carrots need their greens to be upright, so take this concern seriously. Fortunately, there's an easy fix: simply "hill up" the greens by adding more soil around them and forming a larger base for them until they stand up. This tactic will also be helpful if carrot roots begin to appear too early, which can turn them green and bitter.

### Whitish Spots Are Forming on the Leaves and/or Soil

Frequent watering helps carrot growth, but it can also lead to mildew formation, which is what's happening here. Carrots, in particular, can be prone to mildew because they need ample water, and mildew often forms in a field or garden after too much rain. All is not lost, though. You can either use a commercial antifungal spray or apply a natural spray made up of these ingredients:

- 1 gallon of lukewarm water

- 1 Tbsp baking soda

- 2 to 3 drops of mild liquid soap (such as Dr. Bronner's)

- 1 tsp of olive oil

Mix well, and put in a spray bottle, then mist the plant daily. The solution helps to tweak the pH balance on the surface of the leaves, which usually takes care of the mildew.

# Harvesting and Preservation

## Getting Ready to Pick

Much like radishes, carrots indicate when they're ready to be picked by "shouldering," which means they push up out of the soil with part of the vegetable showing. This is hugely helpful in determining the size of the carrot.

If they aren't shouldering, gently dig into the soil with your finger until you can feel the top of the carrot, and this will give you an idea of the size. If they're not ready, just replace whatever soil you've moved and check again in a few days. In some cases, carrots will shoulder too early and begin to turn green, and if this happens, hill up the soil around the carrot to prevent more of the carrot being exposed.

Once they're ready for harvest, just take hold at the base of the leaves and give a firm tug. It doesn't take much effort, and they tend to come out easily.

## Storage Considerations

Carrots are well suited for long term storage, and they're a staple in many root cellars. The trick to storing them for weeks or even months is to refrain from washing them after harvest (save the scrub up for right before eating), since that can hasten spoilage. Also, if possible, arrange carrots so they aren't touching each other.

Whether you plan on short-term storage or a longer timeframe, remove the greens because they draw moisture away from the carrot and make the vegetable dry out faster. If properly stored, carrots can keep in the refrigerator up to three months.

*Carrots, like other root vegetables, tend to "shoulder" as they mature, and this makes it easy to judge when they are ready for picking. A shoulder of about 1-inch diameter is a good range for most carrots.*

# Kale and Chard

Much like lettuce, kale and chard are some of my favorite foods, and we grow plenty of them at the farm. Because they're such nutritional power-houses, I also like growing them on a smaller level at home, and I tend to use a fair amount of both in my microgreens mixes.

Until we started farming, I wasn't aware of the breadth of varieties for both these hearty greens, since I usually saw only a few kinds of kale and two types of chard at my local co-op. But you'll find a beautiful array of options once you start traveling down the rabbit hole of seed catalog perusal, believe me.

Each has unique benefits when it comes to health, but they tend to have the same requirements for growing, which is why they're grouped here. Technically, I could also put collards into this category, and feel free to use these instructions if you want to add those to your indoor growing mix. But I've found that collards do best when they're allowed to grow to full maturity, which usually means leaves the size of dinner plates. That requires a mighty big container for an indoor space, but if you have a large planter and you're a collards lover, that would be a nice option for your indoor edibles mix.

*All chard is attractive to a certain extent, but red chard (shown) and rainbow chard are particularly colorful and striking.*

# Get Ready

## Good Varieties for Indoor Growing

Kale, in particular, comes in an array of varieties, with some growing a few feet tall and others remaining as squat and sturdy as cabbage. For indoor growing, I tend to like the smaller varieties that still offer plenty of texture and color variation, like these:

'Red Russian' kale

Curly kale stays small; look for dwarf varieties

*'Dwarf Blue Curled':* There are a few "dwarf" varieties, and they work well inside because they don't take up too much space. This one, which takes about 55 days from seed to harvest, has a dense frill and a very slight bluish tint.

*'Red Russian':* Thanks to its mild flavor, this is a nice choice for adding raw to salads. The color is stunning, with purple stems and dark reddish green leaves.

'Lacinato' (dinosaur kale)

*'Lacinato':* Also known as Dinosaur, this variety seems to inspire intense devotion in its fans. I like it for the super durable leaves, which hold up nicely during cooking.

With chard, I love the 'Rainbow' variety (also called 'Bright Lights') because the colors are so gorgeous. From neon yellow to delicate pink, the stems of the 'Rainbow' hold their colors even after cooking, making them a nice addition to dishes. Other varieties that are also good for indoor growing include:

*'Peppermint':* I wish this was a minty-tasting chard because the weirdness of that combination would be worth trying. But in actuality, it's a variety that has pink and white striped stems. If you're harvesting at "baby" stage, it's only about thirty days from seed to harvest.

'Bright Lights' chard

## Trays, Pots, and Other Containers

Even though you're choosing smaller varieties of kale and chard, the plants still need plenty of room to stretch out, so this might be a good time to utilize a larger planter that you have on hand. I've attempted to seed one chard plant in a medium-sized pot and the results were mediocre; I had much better growth when using a large rectangular pot that fit well underneath my full-spectrum fluorescents.

## Prep Work

When planting, aim for late spring or early fall. It's always tempting to plant in the winter so that you'll have chard in February, but even with plenty of light and warmth, the results tend to be spindly, and the germination is sluggish. Instead, planting in the early fall will allow the vegetables to establish before the cooler weather hits in earnest.

In terms of soil, standard indoor potting mix will work, and just make sure that there's adequate drainage. You can also blend some compost into the soil for added nutrition.

Yellow chard

*Orange or Yellow:* Unlike other types of vegetables, chard doesn't tend to have catchy names, apart from 'Fordhook Giant' or 'Silverado'; instead, they're identified by stem and vein color, but that can actually be helpful when choosing, since you can grow only orange, pink, red, or yellow varieties.

# Planting and Care

## First Steps

In order to get a nice amount of moisture throughout the soil before planting, mix your soil with some water until it has a damp, but not soaked, consistency. You can determine adequate water level by grabbing a handful and squeezing—if water runs in a steady stream, then it's too wet and you need to add more dry soil. If just a drop or two of water comes out, that's perfect.

Put the soil in your container (if you've mixed it elsewhere), leaving about an inch on top from the edge of the container.

Another nice trick for getting more moisture into the plants is to soak your seeds in lukewarm water for a few hours before planting. This will help boost the germination process, particularly in houses that tend to be dry.

Sow two or three seeds in holes about half an inch deep and at least 5 inches apart. Cover lightly with soil, and then mist the soil until it's moist but not saturated. Put the container in the sun or under a grow light.

*Chard and kale both require good airflow, so don't pack them too tightly together and put a small fan out to create breeze around them.*

## Maintaining Growth

Here are some tips on getting your kale and chard from seed to harvest:

- If growth seems slow, add some nutrients by mulching the top of the plants with compost, coir, even grass clippings or dried leaves. You can also add natural fertilizer such as a kelp blend.

- Water when needed. Check the soil daily to see if it seems dry, and water if necessary. Don't water automatically, since you don't want the roots to stay damp continually, which can lead to rot.

*Soak kale or chard seeds in lukewarm water for a couple of hours before planting.*

Maintain good airflow. Kale and chard are high in antioxidants, but only because they have to contend with wind when they're outside, which creates stronger plant defenses. Mimic that natural environment by increasing airflow around the plants, and changing the "wind" direction occasionally. You don't want airflow to be so strong that it's making the plants flatten, but it should be robust enough that the plants need to work against it in order to grow heartier.

## Troubleshooting

Some common problems with kale and chard and potential solutions:

### Germination Seems to Be Taking a Long Time

There can be several reasons why germination is sluggish for kale or chard. Your house might be too cold or too hot, or the plant could be situated too far from a light source. Try moving the light closer to the container, and also re-mist the soil with water, then cover it with plastic wrap to create a mini-greenhouse environment.

### The Plants Seem to Be Crowding Each Other

Thin out the weaker plants by either plucking them from the soil or by cutting the small leaves close to the top of the soil. The stronger plants will then have more room to grow, and you can eat the leaves you've removed.

# Harvesting and Preservation

## Getting Ready to Pick

Kale and chard can be harvested at any stage of growth, from petite green up to full leaf. I tend to harvest my indoor varieties when the leaves are about the size of my palm, as opposed to the size of my forearm, which is when I harvest varieties grown outdoors.

When harvesting, cut low on the plant to get some of the stem, since this part can be chopped up and used in stir fries, or quickly steamed to maintain crunchiness. Be sure to avoid the inner parts of the plant when harvesting, since this will be where new growth occurs, and snipping these leaves will hasten the plant's end.

## Storage Considerations

Like lettuce, I tend to harvest kale and chard only when I'm about to eat it or juice it. If you need to store the leaves for a few days, though, simply put them in a plastic bag in your fridge's crisper drawer, or in a glass container if you're using the main part of the fridge. Since the higher shelves of a refrigerator aren't as cold as the lowest shelf, I store them there so they can stay crisp for longer.

# Spinach

Technically, spinach can come under the same category as kale and chard, but I decided to tweeze it out into its own section because it has a few unique requirements. Also, to emphasize its awesomeness.

When I was a kid, spinach was my most hated vegetable, by a huge degree. This may have been because we never had Brussels sprouts, but mostly it was because the spinach we had was the canned, clumpy kind that smelled like old, dirty underwear. There was an acrid tang to the aroma that transferred into the taste, and there was literally no amount of money I could have been promised in exchange for eating it.

Then, around the time I stopped eating snack food as a meal, I found out about real spinach—the kind with tender leaves, and it looked like a plant instead of a greenish clot. These days, I love growing spinach in part because I'm still amused by my dietary turnaround. The plant is often so hearty that it can produce for months, and the flavor is delicate, especially when steamed. So, I grow it indoors during the winter for its nutritional prowess, but also as a nod to that super picky kid—I have so much time I have to make up for with spinach.

*Spinach is adaptable and tough and can be moved around as you need. Avoid too much direct sunlight and any rooms that are over 75 degrees Fahrenheit for extended periods of time. Try putting some on top of the refrigerator and reserve the more visible spots for plants that need more maintenance or are prettier.*

# Get Ready

## Good Varieties for Indoor Growing

There are numerous varieties of spinach, but those designated as "baby leaf" tend to work best indoors because they're meant to be harvested at a small size. The leaves will tend to have a mild flavor with a nice bit of crunch that pairs well with salad greens. Here are a few varieties to consider:

Baby leaf spinach ('Catalina')

'Catalina': This kind is the most popular for indoor growing, and for good reason. The deep green oval leaves are very flavorful and the variety is designed to grow quickly.

Red varieties of spinach

'Red Cardinal': This is a fun option for a color change, since it has red veins and stems and dark green leaves. With this one, the good news is that it's designed to be harvested at baby leaf stage, but the bad news is that it tends to have a shorter lifecycle than many other varieties.

'Red Kitten': If you want the red veins but crave a longer lifespan, go for this awesomely named variety; keep in mind, though, that it grows somewhat slower than 'Red Cardinal'.

'Emu' spinach

'Emu': Unlike varieties like 'Catalina', with oval leaves, this one has pointed leaves, which make for a nice contrast.

## Trays, Pots, and Other Containers

Spinach has a shallow root system, so you can use smaller pots if that's handy for you. They should be at least 6 inches deep, so in some cases, you can use the leftover trays that you might have used for microgreens or pea shoots, just be sure to scrub them thoroughly and let them dry completely before you replant in them.

I tend to like rectangular pots because they fit well on my shelves under my fluorescent lights, but I've seen many configurations for spinach, from growing a single plant in an old plastic tub to combining spinach with herbs in a larger container.

*(BELOW) The plastic oblong planter is perfect for a nice little indoor spinach patch.*

## Prep Work

Spinach appreciates cooler weather, so aim for planting in early fall if possible so that the plants can establish before any major temperature drops. At our farm, we experimented with planting spinach in October and then covering it with reemay, a landscape fabric that lets light in but keeps bugs out. Even in a tough Minnesota winter, the spinach did great and we were able to harvest in early spring, after it "woke up" from its hibernation. Similarly, you can keep it going in your house during the winter if you live in a chilly climate, but it's best to get it germinating and growing before any major temperature variations.

In terms of soil, standard indoor potting mix will work, and just make sure that there's adequate drainage by adding some sand, coir, or vermiculite. You can also blend some compost into the soil for added nutrition.

# Planting and Care

## First Steps

In order to get a nice amount of moisture throughout the soil before planting, mix your soil with some water until it has a damp, but not soaked, consistency. You can determine adequate water level by grabbing a handful and squeezing—if water runs in a steady stream, then it's too wet and you need to add more dry soil. If just a drop or two of water comes out, that's perfect.

Put the soil in your container (if you've mixed it elsewhere), leaving about an inch on top from the edge of the container.

Sow two or three seeds in holes about half an inch deep and at least 5 to 8 inches apart. Cover lightly with soil, and then mist the soil until its moist but not saturated. Put the container in the sun or under a grow light.

You should begin to see germination in about seven to fourteen days, depending on variety and growing conditions.

## Maintaining Growth

Here are some tips on getting your spinach from seed to harvest:

- Watch the temperature. Spinach doesn't do well in very warm conditions, so if your house is over 75 degrees, move the plant to a cooler spot.

- Water when needed. Check the soil daily to see if it seems dry, and water if necessary. Don't water automatically, since you don't want the roots to stay damp continually, which can lead to rot.

*Poke holes about ½ inch deep for spinach seeds and plant two or three seeds per holes.*

*Instead of planting spinach from seed in your containers, you can start it outdoors or in a greenhouse and then transplant the small spinach plants when they are strong enough.*

*If your spinach flowers that means it is bolting and it may already be too late to harvest it before the leaves become highly bitter. Some types of spinach, like the Malabar seen here, do have edible flowers.*

## Troubleshooting

Some common problems with spinach and potential solutions:

### The Spinach Just Started Growing, and Now It Has a Flower in the Center

This is called "bolting" and it's a sign that the plant is nearing the end of its life. Most likely, the timeframe for your spinach got significantly shortened because the plant got too hot. If other plants aren't bolting yet, move the planter to a cooler place and be sure to pick off any tall stems that look like they're about to flower. You can also detect signs of bolting when the spinach leaves taste more bitter than they did previously.

### Germination Seems Uneven

Sometimes, several spinach plants in the same pot will germinate at different times, leaving "gaps" in the soil. Simply plant more seeds in those spaces, since they can catch up quickly. Since germination is usually fairly quick, they won't be far behind those that were planted the week before.

### Leaves Are Wilted, Bruised, or Yellowing

Most likely, this is a sign of overwatering or of not enough nutrients. Try to "dry out" the plant for a day or two and when you resume watering, place the pot in a tray or sink filled with a few inches of water and remove it when you can see the soil on top begin to dampen. Sometimes, spinach pots can develop a "dead zone" inside where water flows around that area inside the pot, leaving it overly dry even though the soil on top shows adequate water.

# Harvesting and Preservation

## Getting Ready to Pick

Spinach can be harvested at any stage of growth, from petite green up to full leaf. If you've planted a baby leaf variety, harvest when the leaf is about 6 inches long.

When harvesting, cut low on the plant to get some of the stem, and be sure to avoid the inner parts of the plant when harvesting, since this will be where new growth occurs, and snipping these leaves will hasten the plant's end.

## Storage Considerations

Spinach will keep for about a week in the refrigerator if properly stored. Don't wash the leaves until you're about to eat them, and wrap the spinach in plastic wrap, squeezing out as much air as you can. Unfortunately, cooked spinach doesn't store very well—it tends to turn into the glop I remember as a kid—but the plastic wrap tactic will at least keep your fresh spinach ready for cooking for a few days, if not longer.

*Spinach leaves should be snipped off, not torn, near the base the stem. Take the outer leaves first to give the inner leaves a chance to grow.*

# Hot Peppers

In my indoor growing ventures, I've had limited success with bell peppers, mainly because they require a higher degree of warmth and more light than everything else I'm growing.

As you may find with your own indoor gardening strategies, having one plant that demands more of your attention and is kind of a space-and-resources hog can make your gardening feel like it's taking up more time than you want to give. Personally, I like to set my schedule for indoor gardening—more hours in the winter, barely any in the summer—and having one vegetable that throws it all off is a primary candidate for getting booted from the grow space.

Hot peppers, on the other hand, are a different story. Some varieties grow to a nice, manageable size for a container, and although they do have unique needs when it comes to heat and watering, they're easily maintained.

*Hot chile pepper plants generally fare well indoors and are well proportioned for most indoor kitchen gardens.*

# Get Ready

## Good Varieties for Indoor Growing

Although it might seem that hot pepper plants have an especially shallow root system—they're dishearteningly easy to knock over or pull out of the ground by accident—there are several varieties that rely on extending their roots fairly deep into the soil. That's why it's particularly important to choose a variety that's better suited to indoor growing, unless you want to invest in larger containers. These tend to work well for an indoor garden:

Habanero pepper

'Habanero': With a characteristic orange color and a very pungent aroma, this is one of the hottest peppers you can grow. If you're sensitive to capsaicin, which gives peppers their heat, you may want to use gloves when harvesting and handling this one.

'Firecracker' chile pepper

'Hungarian Hot Wax' pepper

'Firecracker': With this variety, peppers grow straight upward rather than hanging down, creating a firework-type of effect. Peppers turn from green to yellow to red while maturing.

'Hungarian Hot Wax': Particularly good for making pickled peppers, this variety produces robust fruit that changes from yellow to red, and is spicy but not my-mouth-is-on-fire hot.

## Trays, Pots, and Other Containers

Like many other plants, hot peppers will do better in plastic than in terra cotta pots, because the clay will dry out the soil mix faster than the plastic. Considering that the peppers will require more heat and moisture than many other plants you might be growing, using terra cotta will thwart your efforts.

When selecting a planter, make sure that it's thoroughly cleaned if you've grown plants in it previously. Even with old pots, it's possible to transfer diseases onto the new crop.

## Prep Work

All indoor plants need attention when it comes to water, light, and heat, but hot peppers are especially finicky about these three factors. Make sure you have either a sunny spot by a south-facing window or adequate grow lights. The lights should be positioned only about 6 inches above the plant while it germinates, and should be left on for fourteen to sixteen hours per day.

In terms of soil, standard indoor potting mix will work, and just make sure that there's adequate drainage by adding some sand, coir, or vermiculite. You can also blend some compost into the soil for added nutrition.

*Glazed ceramic pots work very well with pepper plants from a design standpoint.*

# Planting and Care

## First Steps

With hot peppers, you can speed up germination time by tweaking the planting process, starting with getting moisture into the seeds. Follow these steps:

- Keep the container or bag in an area that's warm but not hot. In my house, this is on a shelf under our butcher block that's near the stove. If it's cold in the house, I might use a germination mat and put the container on that.

- Check the seeds after a few days, and if they seem puffed up then you're ready to plant. Sometimes they might even be starting to sprout, which is a great head start.

In order to get a nice amount of moisture throughout the soil before planting, mix your soil with some water until it has a damp, but not soaked, consistency. You can determine adequate water level by grabbing a handful and squeezing—if water runs in a steady stream, then it's too wet and you need to add more dry soil. If just a drop or two of water comes out, that's perfect.

Put the soil in your container (if you've mixed it elsewhere), leaving about an inch on top from the edge of the container.

Sow each seed about half an inch deep and at least 2 inches apart. Cover lightly with compost, and then mist the soil until it's moist but not saturated. Put the container in the sun or under a grow light.

You should begin to see germination in about one to six weeks, depending on variety and growing conditions.

Dampen two paper towels and place your seeds in a single layer across one of them, putting the other dampened paper towel on top.

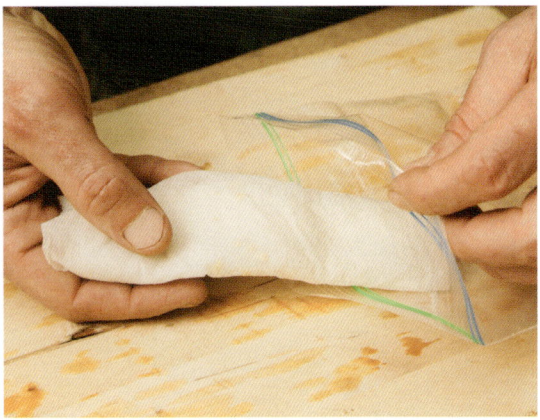

Put the seeds and paper towels into a plastic bag that can be sealed or a plastic container with a tight-fitting lid.

## Maintaining Growth

Here are some tips on getting your peppers from seed to harvest:

- Water thoroughly after the peppers germinate. Peppers thrive with regular watering, and you should mist or water when the soil is just beginning to look dry.

- Check airflow so that there's circulation, but not an indoor wind tunnel. For a few hours a day, turn on a small fan in the area, set on low. This will help to mimic outdoor conditions, but you don't want to bring down the temperature of the room too much.

## Troubleshooting

Some common problems with hot peppers and potential solutions:

### Growth Seems Sluggish

Check your seed packet for seed-to-harvest maturity dates first, since peppers can vary widely in terms of growth timeframes. If it's off track, make sure that you're not placing the plants near areas that experience temperature fluctuations, such as a countertop under a heating vent or an air conditioning vent. Even placing the plant near an outside door that brings in cold air occasionally can cause the plant to slow its growth.

### Germination Isn't Working, Even With the Paper Towel Method

Sometimes, pepper seeds can be fussy. In that case, it might be time to invest in a heat mat, which can be purchased at any garden store. These rectangular mats look like tiny heating pads, but they don't get as hot. If I'm experiencing slow germination times, especially in the winter, I place the pot on this type of mat and wait a few days—usually, the mat will heat the soil enough to foster growth. The heat is so minimal that you can put the mat on any surface, too; I usually put mine on a wooden counter, without any fear of scorching it.

# Harvesting and Preservation

## Getting Ready to Pick

Picking chili peppers off your own indoor plant has got to rank as one of the happiest experiences of home gardening. Depending on variety, you're likely looking at a cheery, brightly colored pepper that will taste amazing. Keep in mind that peppers will start off as green and then will gradually transition to a color if that's part of the variety you've chosen. If you're growing a type that stays green, you can check for "doneness" by gently squeezing the pepper—if it's rock solid, then it's not ready, but if it has some give to it, then you can harvest.

The best way to harvest is to clip the pepper close to the stem, rather than plucking them off, which can sometimes yank the plant along with it.

## Storage Considerations

Peppers will store nicely in the refrigerator for at least a week, especially if you don't wash them first. They also dry exceedingly well with a food dehydrator or in the oven. When we have a bumper crop of peppers, I dry a bunch of them and then vacuum seal them to use later for chili or stew.

*Red chili peppers, and many other types of peppers, dry readily in many environments, including most kitchens. They also look great, so don't be afraid to have some fun using them in your décor.*

*When red chili peppers are bright red, snip them as they are needed with scissors. Avoid plucking them off the plant.*

# Potatoes

In many cases, potatoes grow so well outside and store so beautifully that growing them indoors is more of a jaunty experiment than a practical venture. However, I'm a fan of taking on any kind of experiment, especially if it's jaunty.

Sometimes, indoor growing can be an advantage for potatoes, too. In our fields, we have to fight potato bugs on a yearly basis, and because we're big believers in organic practices and natural pest management control, that tends to mean that I squish them instead of spray them. Bright orange and fat as woodticks, potato bugs leave me with stained hands and a slightly nauseated feeling. But at home, I can just enjoy watching the potato plant's progress without the dread of "squashing day."

Also, there's a certain satisfaction that comes with knowing you're growing potatoes in your base-ment. It feels like being a modern pioneer.

*Potatoes, believe it or not, can be grown in pots indoors.*

# Get Ready

## Good Varieties for In door Growing

Farmers and gardeners rely on "seed potatoes" for planting, but that term doesn't mean that these potatoes are any different than the type you eat. In fact, many farmers (including us) put aside their best potatoes during harvest—which often means the ones with the most eyes and seem especially firm and hearty—and use those for the next planting.

Theoretically, you could also go into the grocery store and choose a bag of potatoes and call those your seed stock. However, this is usually a less-ideal solution because most potatoes, even many that are organic, are treated with anti-sprouting agents that extend their shelf life. Those supermarket potatoes might sprout at some point, but to grow the best potatoes possible, buy some potatoes that were specifically developed to be used as seed.

*Because of their small size, fingerling potatoes can be a good indoor growing choice.*

When researching your options, here are a couple factors to consider:

- Keep in mind that depending on your state, you may not be able to order the potatoes during certain months. This is because seed potatoes rot quickly if they're allowed to freeze, so seed companies won't ship them to colder climates if there's a chance that they'll get spoiled.

- Seed potatoes can run out once summer approaches. This makes sense, since most of these potatoes are harvested in early October and then are ordered by farmers and gardeners over the winter and early spring. By June, and sometimes even by April, the pickings get slim. So, plan ahead.

- One pound of seed potatoes will result in about 10 to 15 pounds of potatoes. It doesn't take much to make a pound of potatoes, so when ordering, lean toward the conservative side.

- Much like other indoor crops, choosing smaller varieties tends to work well. I've had success with fingerling varieties like 'Russian Banana' and 'French Fingerling', as well as more petite options like 'Mountain Rose', 'Rio Grande', or 'All Blue'.

- Compared to other crops you might be growing indoors, potatoes take a long time, especially if you've got trays of fast-sprouting wheatgrass and microgreens in the mix as comparison. The quickest timeframe you can get is about seventy days, but usually, you're looking at about four months. They don't require much maintenance at all, so it's not like you have to tweak your vacation schedule to grow them, but planning ahead is helpful so you can anticipate harvesting.

## Containers

Here's your chance to use your biggest pot or container. Some people have used 5-gallon buckets with holes drilled in the bottom, or special potato growing bags that are fashioned out of heavy felt that can be easily moved around.

Whatever you choose, make sure that there's adequate drainage options, and that the pot is in an area where the water can drain easily. You'll be using a fair amount of water during the initial watering period and when you do maintenance watering, so placing the pot in a spot where you have easy water access and a drain is helpful—for my setup, the area near my laundry sink in the basement has proved to be ideal.

## Prep Work

The main preparation for seeding potatoes is cutting and sprouting. For some smaller potatoes with only one "eye," this isn't a necessary step, but for the majority of seed potatoes, this strategy helps prevent diseases and ensures a heartier crop. Here are the steps:

Wash potatoes to remove dirt and debris; this is an especially important step if buying non-organic potatoes, because you want to remove any pesticide residue before planting.

- Cut each potato into pieces that include at least one eye per piece. An "eye" is the indentation where a potato will sprout eventually.

- Set potatoes aside for at least a week before planting. The cut side should be allowed to toughen up, in a process also called scabbing. The eyes will begin to sprout, and this is a sign that the potato is ready to plant.

- Alternately, you can employ a method that's often used for teaching kids about plant growth: put four toothpicks around the potato chunk, about halfway between top and bottom. Put the potato eye-side down in a glass jar of water, with the toothpicks preventing the potato from submerging completely. Put the jar in a sunny spot, and change the water if it begins to look cloudy. Roots will sprout in about a week. This method can speed up sprouting if that's what you want, but you won't get the scabbing that helps with disease control.

Wash the seedling potatoes well.

Cut the seedlings into small pieces.

In terms of soil, you can use indoor potting mix for vegetables, and mix in vermiculite or sand to help with drainage. Resist any urge to include outdoor garden soil in the mix, even if it's rich with compost and other nutrients. You don't want to take the chance that you'll bring in pests from outdoors.

# Planting and Care

## First Steps

Fill the pot or container about a third full of your soil mix. As the plant grows, you'll need to keep adding soil, so it's best to stay minimal with the amount at the beginning, even if it looks like the pot is mainly empty.

Place the potato chunks about 6 inches apart, and 3 inches deep, with root side down. Because the plants will need room to grow, it's best to put the seed potatoes at least 6 inches from the edge of the pot as well.

Cover the potatoes with about 3 inches of soil, and water them thoroughly, until water begins running out of the drainage holes in the bottom.

## Maintaining Growth

When the plant has grown about 6 inches above the surface (about one to three weeks after planting) add more soil to the pot without completely covering the new growth. When the stem reaches the level of the top of your pot, add soil again and hill it around each stem. This helps to ensure that the potatoes will grow more deeply under the soil and won't be subjected to sunlight, which would cause them to turn green.

Green potatoes, while they sound colorful and charming, are mildly toxic because they contain a substance called solanine, which occurs when the plant feels threatened. The length of exposure to light, and the intensity of the light, can raise the level of solanine, with the highest concentration located in the potato's skin. This whole process can happen when the potato grows too close to the soil surface, or is stored long-term in an area with even low light levels. If you don't spot any green tint, you'll know when solanine is present because your potatoes have a bitter aftertaste. Prevent the issue by being diligent about hilling up the potatoes so they're well protected from light.

Keep the container well watered, but not soaking. The soil doesn't have to be particularly damp, but you should have the potatoes on a consistent watering schedule to foster more growth.

## Troubleshooting

Some common problems with potatoes and potential solutions:

### Plants Don't Come Up After Planting Seed Pieces

If you've taken a chance on using supermarket potatoes instead of certified seed potatoes, you may be thwarted in your growing. Also, if you planted the seed pieces before they sprout, they'll take longer to fully establish.

### There Are Cute Little Yellow and Black Beetles on the Leaves

Evil takes many forms, and this is one of them. These are Colorado potato beetles, and you need to pick them off before they destroy your plants.

You can squish them if you're feeling resentful, or simply throw them into some soapy water, where they'll expire. To make sure that you've gotten the eggs as well, cut up some basil leaves and let them steep in about a cup of water for a few days. Strain, put in a mister and thoroughly spray the leaves.

### The Plants Are Staying Green Even Though It's Past the Maturity Date

This is probably the only instance in growing when you want dead, dry leaves instead of vibrant green ones. If this is happening, then it's likely that your grow space is too warm. Potatoes do best with cool nights of below 55 degrees, and even if you can't achieve that in your house, move the potatoes to the coolest spot in the house to see if it helps.

# Harvesting and Preservation

## Getting Ready to Pick

You can tell when potatoes are ready because there are usually very small-looking potatoes on the vines/stems, and the stems themselves will dry up and look terrible. The stem dies off as part of the maturation process, and it's about two to three weeks after they shrivel up that you can harvest the mature potatoes.

To dig them up, you can simply pull up on the stem and you'll see a string of potatoes, as well as the original seed potato, also called the "mother." This potato chunk will be soft and pretty gross and should be discarded. But the others can be collected and stored. Be sure to dig through the soil to make sure you haven't missed any potatoes that may have dropped off during harvest.

## Storage Considerations

If you plan to store potatoes for quite a while, place them in a dry spot out of direct sun, like in a shady patio or a garage. This will help them dry out before storage. Then place them in a cool, dry, dark place like a basement; usually, a mesh or burlap sack is ideal because it allows the vegetables to "breathe" while being stored.

No matter how long you plan on storing them, refrain from washing them between harvest and storage, and wash them only just before preparing them for cooking. This will keep the potatoes from potential rot while in storage.

I've saved tomatoes for last because, in many cases, they feel like an extra credit project. Some gardeners have insisted to me that you simply can't grow tomatoes—viable, tasty ones, anyway—indoors because much of what they need is outdoors. For example: bees.

Pollination through the use of bees provides a major boost. After all, tomatoes are a fruit, and like many other fruits, they rely on cross pollination to increase the quality and abundance of the harvest. But you can game the pollination process (more on that later) if you're truly dedicated to indoor tomatoes.

Usually, even though I am ridiculously in love with tomatoes, I tend to leave them out of my indoor garden because I find them fussy. But then, when I bring them back in, I do have that moment of elation that comes with seeing a tiny green tomato that will eventually transition into a bright little treat.

# Tomatoes

I've saved tomatoes for last because, in many cases, they feel like an extra credit project. Some gardeners have insisted to me that you simply can't grow tomatoes—viable, tasty ones, anyway—indoors because much of what they need is outdoors. For example: bees.

Pollination through the use of bees provides a major boost. After all, tomatoes are a fruit, and like many other fruits, they rely on cross pollination to increase the quality and abundance of the harvest. But you can game the pollination process (more on that later) if you're truly dedicated to indoor tomatoes.

Usually, even though I am ridiculously in love with tomatoes, I tend to leave them out of my indoor garden because I find them fussy. But then, when I bring them back in, I do have that moment of elation that comes with seeing a tiny green tomato that will eventually transition into a bright little treat.

*Smaller tomatoes, such as cherry tomatoes, will do fine indoors if they get suitable amounts of light, wamth, moisture, and air movement.*

# Get Ready

## Good Varieties for Indoor Growing

I'd like to say that I've grown palm-sized, juicy slicer tomatoes that dangle like coconuts in my kitchen, even with the snow going sideways outside. But my attempts at larger tomatoes have sputtered, so instead I stick with what I've found the most successful: cherry tomatoes and small varieties. Here are some favorites:

- 'Yellow Pear': This is a cheery, bright yellow heirloom variety that makes for a nice contrast with red and orange cherry tomatoes; plus, its pear shape is delightful.

- 'Pink Ping Pong': named for the size of the fruit, this cherry tomato doesn't actually get pink, but does have a slightly lighter reddish color.

- 'Tommy Toe': Originating from the Ozark mountains, this heirloom has high disease resistance and tends to yield copious amounts of tomatoes.

- 'Alaska': This is a nice pick for cooler climates, and it hails from Russia originally; an early producer, the tomato plant yields larger cherry tomatoes.

- 'Cream Sausage': A quirky tomato that's fun to grow, because it's bushy and so doesn't need trellising; also, the tomatoes are long and come to a point at the end, so they resemble peppers.

There are so many more—tomatoes, particularly heirlooms, come in hundreds of varieties, each with their own distinctive characteristics.

One important note is to look at determinate versus indeterminate varieties. The former means that the plant has been developed to reach a specific height and number of tomatoes per plant. They tend to be bushier and therefore don't need as much trellising or staking.

One drawback to determinates is that the tomatoes all ripen at the same time, so you don't have the ongoing fruit development that you'd see with indeterminates, which produce fruit until they're killed by frost. Indeterminates are like the grab bag of the tomato world—you don't know what you're going to get in terms of height, fruit amount, or life cycle, but that's part of the fun.

'Yellow pear' tomato

## Containers and Trellis Systems

Tomatoes do well with plenty of space for root growth, so opt for large containers and pots. These bigger containers also give you an opportunity to do more with staking and trellising, which is important for healthy plant growth. I've seen some successful grow systems that use 5-gallon buckets, which are handy because they can be easily moved. Any smaller than that size and you'll likely cramp the roots as they're trying to establish.

At this point, do I even have to mention the importance of drainage? Or of drilling holes in those 5-gallon buckets? If my editor would have allowed it, I may have been tempted to title this book *The Importance of Drainage: Indoor Growing for People Who Have Appropriate Pots for the Task.* Seriously, though, by now I'm sure you get it—drainage is awesome.

For a trellis system, I've seen all kinds of strategies, from a yardstick stuck into the soil with twine holding the plant to an elaborate grid-type plastic trellis that allowed for intensive support and attached to a "self-watering" pot.

Many people on a budget (like myself) do well with the standard tomato plant cage, which is often under $10 (and sometimes under $5) at garden stores. This triangular or round cage is about 3 to 4 feet high, made of wire, and can easily be stuck into any planter. Because it has several concentric rings, you can weave your growing tomato stems inside of it and tie them gently to the rings or sides.

*With its compact cage and self-watering pot, this small patio tomato is all set to move indoors.*

## Prep Work

Unless you have a greenhouse space that you can outfit with some serious grow lights, it's best to utilize natural sunlight for tomatoes. This usually means finding the biggest south-facing window in your house and getting the tomatoes as close to full sun as possible. If you live in more southern climates, you may want to opt for an east-facing or even north-facing window.

If it's the middle of winter and you're barely getting enough sun yourself, then supplement the anemic natural light with some full-spectrum fluorescents. Dedicate at least two of these to your tomatoes, and place them close to the plants--usually no farther than a foot away.

In terms of soil, a good indoor potting soil will work, but be sure to increase the nutrient density with compost or other fertilizer. Also, help with drainage by putting some rocks or pebbles in the bottom of the pot so water won't sit inside the soil, leading to root rot.

*In-home tomato plants can be started indoors too.*

# Planting and Care

## First Steps

You have a couple choices with how to establish tomato plants. You can start them from seed, but keep in mind that this tends to take quite a while and indeterminate varieties will only begin flowering when they're at least 3 feet high.

If you do begin from seed, consider using much smaller pots for the first stages of germination, since this will help to create a "root ball" and lead to a healthier, more robust plant. Simply put a seed into a small pot, about 2 inches deep, cover with soil, and water consistently until you see growth.

As the plant begins to expand, transplant to a medium-sized container that will support the growth and put a small stick or ruler in the soil if you need to do a mini-trellis. After more growth, do one more transplant session to the larger container where you plan to have the plant for the rest of its life cycle. Each time you transplant be sure to water thoroughly—transplanting is often traumatic to a plant and it requires extra watering and care to recover from the shock.

*Tomato seedlings are cheap and reliable, but they are only available for a short time in the spring in most areas.*

You can also start your indoor tomatoes from cuttings, using outdoor plants at the end of the summer. Just snip a branch from a variety in your garden (or, with permission, someone else's garden) and strip off all the leaves except one on top. Put the stem in a jar of water and place in a window that gets plenty of sun. Roots will begin to form from the lower part of the stem, and once you see a few inches of root growth, put the stems in the large container, burying most of them but leaving the top leaves exposed.

Finally, you can simply buy transplants from a garden center. I promise, this isn't cheating. I like to call it a "head start" instead. Tomatoes take time, fertilization, water, and care to grow from tiny sprout to firmly established plant, and when I have a ton of other plants growing and just don't want to take the time and effort to nurture tomato plants along, getting some healthy, beautiful transplants from my locally owned garden center seems like a perfect strategy.

## Maintaining Growth

As the plants begin establishing and getting larger, here are some strategies for fostering more growth:

- Hand pollination by gently shaking the plant's flowers daily; some people go beyond this tactic and use a paintbrush to "vibrate" the flowers. Either method is helpful, because it mimics the action of a bee or other pollinator. It's also fun to research different indoor pollination methods because people are very creative and come up with some surprising tactics, like using ostrich feathers, tuning forks set for middle C (seriously), or an electric shaver. Many people report

particularly good results with an electric toothbrush, using the handle instead of the brush, because it shakes the flower as it dislodges the pollen.

- Keep proper airflow. Also an aid to pollination, and a generally good health booster is air circulation, so make sure that your growing space has a fan directed toward it. I use a rotating fan since that replicates that occasional breezes that would occur outside, as opposed to setting a fan to blow directly on the plant at all times, which could cause stress eventually.

- Create a natural day and night. Although tomatoes love sun, they also need to rest in order to grow properly, much like any other plant. So if you're utilizing growing lights, be sure to simulate a day by turning lights on at sunrise and off at sunset. If you're in northern climates during the winter, you may want to create a longer day for the tomatoes by extending your "day" for them by an extra few hours. This is where a timer can come in handy.

- Water consistently. Just like they love sun, tomatoes also appreciate regular watering. The soil in your container should be damp but not soaked, and it's likely that you'll have to water every few days, depending on the dryness of your house's air.

- Watch for insects like aphids, spider mites, and whiteflies. These are common in indoor spaces and can cause serious damage to your tomatoes. Either place sticky insect traps near the tomatoes or employ a soapy spray that consists of mild soap (like Dr. Bronner's) diluted in water and put into a spray bottle.

# Troubleshooting

Some common problems with tomatoes and potential solutions:

## Leaves Look Slightly Yellowed and Growth Seems Slow

When tomatoes seem to be struggling, they rebound nicely with some fertilizer application, particularly kelp and bone meal. Tomatoes are "heavy feeders," which means they need plenty of nutrients, and even with compost mixed into your soil, it's likely you'll have to add some fertilization along the way. When leaves begin to yellow, it's often a major sign of either nutrient deficiency or overwatering.

## Tomatoes Have Black Areas on the Bottom That Look Burned

This is called blossom-end rot, and although it looks fairly serious, it's not a disease and can be controlled. The condition is sometimes caused by a calcium deficiency brought on by an uneven watering schedule. In the field, we see this problem when it rains heavily and then we're hit by drought. Another factor might be excess nitrogen, so if you're overfertilizing, you might see this form on several tomatoes. Fortunately, if you catch this early, you can correct the issue by keeping the soil at a consistent moisture level, and also by mulching—just cover the top of the soil with straw, available at any garden store, which helps to keep moisture in the soil.

## Dark Spots on the Leaves

Another major thorn in the side of tomato growers is this fungal problem, known as "blight." For some home gardeners, blight takes out their tomato plants every season, and they try planting in new places to alleviate the issue. When growing indoors, it's easier to spot blight early and try containment methods. First, make sure you're not getting any water on the leaves or touching the plants when they're wet. Don't mist the plant, ever, as this will worsen the problem. Instead, pick off the affected leaves and destroy them. Increase air circulation to keep the leaves as dry as possible. If the issue is extensive, consider using a spray like copper fungicide, which is appropriate for use on organic plants. In addition to early blight, the spray also tackles powdery mildew, black spots, and downy mildew.

## Fruit Developing Cracks

Here's another indication of uneven watering. See the answer above for strategies that will help alleviate the problem.

# Harvesting and Preservation

## Getting Ready to Pick

Nothing could be easier—just pluck the ripe fruit and see if you can resist popping the fruit in your mouth immediately. Ripeness will depend on variety, but expect your tomatoes to go through several color variations as they ripen, from green to vibrant color.

## Storage Considerations

Tomatoes do well on the counter for a few days, especially if you need them to ripen more, which they'll continue to do even off the vine. Just don't put tomatoes in the fridge (unless you've turned them into salsa) because it significantly impacts the texture and flavor of the fruit.

*Picking tomatoes is easy: just twist and pull.*

# Photo Credits

Peter Abell/Alamy: pp. 104 (left)

Peter Anderson/Dorling Kindersley/Getty Images: pp. 98

Lee Avison/Gap Photos/Getty Images: pp. 126

Katie Elzer-Peters: pp. 92

flowerphotos/Alamy: pp. 132

iStock: pp. 33, 51, 53, 58, 65, 125 (left)

Johnny's Selected Seeds: pp. 56 (all), 57 (all), 68, 69, 110 (top), 133

Crystal Liepa: pp. 6, 7, 8, 12, 13, 14, 15, 16, 17, 20, 21, 22, 24, 27 (all), 29, 30, 31 (bottom), 32 (left), 32 (right), 34, 35, 36, 38, 39, 40, 41, 42, 44, 52, 54, 55, 60, 61 (all), 62, 63, 64, 70, 72, 73, 74, 76, 77, 78, 85, 86, 87, 88, 95 (top right, bottom right), 100, 106, 109, 112, 114, 116, 117, 120, 123, 129, 134

J. Paul Moore: pp. 25

Victoria Pearson/The Image Bank/Getty images: pp. 103

Rau+Barber: pp. 4, 48, 50, 90, 97

Shutterstock: pp. 10, 18, 28 (left), 28 (right), 31 (top left, top right), 45, 46, 66, 67, 71, 80, 81, 82, 83, 84, 89 (both), 93, 94, 95 (top), 99, 101, 102, 104 (right), 105, 108, 110 (bottom left, bottom right), 111, 115, 118, 119, 121, 122, 125 (right), 127, 136, 139

Bossy Acres: pp. 19, 59

# Index